Advanced Dynamics and Model-Based Control of Structures and Machines

Hans Irschik
Michael Krommer
Alexander K. Belyaev
Editors

Advanced Dynamics and Model-Based Control of Structures and Machines

Springer

Editors
Hans Irschik
Universität Linz
Institut für Technische Mechanik
Altenbergstr. 69
4040 Linz
Austria
irschik@mechatronik.uni-linz.ac.at

Alexander K. Belyaev
Russian Academy of Sciences
Inst. of Problems in Mech. Eng.
V. O. Bolshoy pr. 61
199178 St. Petersburg
Russia
vice.ipme@gmail.com

Michael Krommer
Universität Linz
Institut für Technische Mechanik
Altenbergerstraße 69
4040 Linz
Austria
michael.krommer@jku.at

SpringerWienNewYork is part of Springer Science + Business Media
springer.at

Typesetting: SPi Global, India

Printed on acid-free paper and chlorine-free bleached paper
SPIN 80063047

With 120 Figures

Library of Congress Control Number: 2011936794

ISBN 978-3-7091-0796-6 e-ISBN 978-3-7091-0797-3
DOI 10.1007/978-3-7091-0797-3
SpringerWienNewYork

Foreword

The Workshop revived a long-lasting and fruitful cooperation between Russian and Austrian researchers in the field of theoretical and applied mechanics. A first step towards this cooperation was made more than 20 years ago when Alexander Belyaev was invited to spend his sabbatical year at the Institute of Rational Mechanics at the Vienna University of Technology headed by Franz Ziegler. When Hans Irschik was promoted to the position of Professor and Director of the Institute of Technical Mechanics at the Johannes Kepler University of Linz in 1991, he invited Alexander Belyaev to teach as a Guest Professor in his Department for 6 years (1993–1999) and thereby established the close relationship between the Russian and Austrian schools of mechanics. Apart from the joint tutorial, scientific and research activities, a number of international conferences were jointly organised by the Russian and Austrian members. It is worth mentioning the *Fourth European Conference on Structural Control* held in St. Petersburg, September 8–12, 2008, co-chaired by Hans Irschik and Rainer Flesch from the Austrian side along with Alexander Belyaev and Dmitry Indeitsev from the Russian side. The cooperation deepened in the joint efforts to make the IUTAM *Symposium on Vibration Analysis of Structures with Uncertainties* held in St. Petersburg, July 5–9, 2009, co-chaired by Alexander Belyaev and Robin Langley, a long lasting success: Franz Ziegler acted as the IUTAM Representative. A sign of appreciation of the Russian-Austrian cooperation is that both Franz Ziegler and Hans Irschik were awarded the title of doctor honoris causa of the St. Petersburg State Polytechnic University.

The recently founded Austrian Center of Competence in Mechatronics, ACCM, in Linz, generously offered the opportunity of pursuing an intense mutual exchange of know-how in this high-tech area of international research interest. The undersigned greatly enjoyed the reunion with distinguished colleagues at this Russia-Austria Joint Workshop on *Advanced Dynamics and Model Based Control of Structures and Machines*, which took place at the Johannes Kepler University in Linz, April 2010. Invited distinguished researchers from Japan, Italy and Taiwan participated in this fruitful Workshop to contribute to the latest achievements in this specific field. The Proceedings of this Workshop will make available up-to-date results to the international community and therefore will stimulate further developments and successful international cooperation in the field of Mechatronics.

St. Petersburg, Vienna
May 2011

Nikita F. Morozov
Franz Ziegler

Preface

This book contains the contributions presented during the *Russia-Austria Joint Workshop on Advanced Dynamics and Model Based Control of Structures and Machines*, which took place at the Johannes Kepler University of Linz, Austria, in April 2010. The workshop aimed at bringing together scientists from Russia and Austria with an outstanding expertise in mechanics and control, with emphasis on the application to advanced structures and machines. The international character of the workshop was deepened by the participation of widely renowned scientists from Italy, Japan and Taiwan. The workshop continued a series of international workshops, which started with a *Japan-Austria Joint Workshop on Mechanics and Model Based Control of Smart Materials and Structures* that took place in Linz, Austria, in September 2008.

This series of workshops is organized within the framework of the Area *Mechanics and Model Based Control* of the *Austrian Center of Competence in Mechatronics (ACCM)*. This peer-reviewed Center served as the Steering Organisation for the workshop series. Mechanics and Model Based Control are rapidly expanding scientific fields and fundamental disciplines of engineering, particularly in Mechatronics. They share demanding mathematical and/or system-theoretic formulations and methods. One challenge in Mechanics and Model Based Control is to use the ever increasing computer power with respect to both, the simulation of complex physical phenomena in mechanics, and the design and real-time implementation of novel control systems. Further challenges follow from the availability of efficient multi-functional materials, so-called smart materials, allowing the design and implementation of new types of actuator/sensor fields and networks. From a strategic point of view, the key objectives of the workshop series are:

- Enabling the interchange of ideas from advanced mechanics of structures and machines, and from control theory.
- Clarification of expectations of researchers in the field of mechanics from advanced control theory and vice versa.

- Development of joint international research proposals and teams.
- Encouragement of collaborations among industry and universities across the borders of the participating countries.

The main topics of the present *Russia-Austria Joint Workshop* were:

- Laminated, composite and functionally graded materials.
- Sensing and actuation.
- Active and passive damping.
- Vibrations and waves.
- Nonlinear control of structures and systems.
- Nano- and micromechanics.

We believe that the workshop will finally result into the creation of research teams with participation not only from Russia and Austria, but also from other countries. Such teams should push the frontiers of advanced dynamics and model based control of structures and machines to new dimensions, resulting into the advanced design of future structures.

The undersigned Editors of the present book, which is entitled *Advanced Dynamics and Model Based Control of Structures and Machines* are happy to present in the following 8 full length papers of presentations from Russia, 12 from Austria, 2 from Italy, 3 from Japan and one paper from Taiwan. It is hoped that these contributions will further stimulate the international research and cooperation in the field.

Linz, St. Petersburg
May 2011

Hans Irschik
Michael Krommer
Alexander Belyaev

Acknowledgements

Support of the *Russia-Austria Joint Workshop on Advanced Dynamics and Model Based Control of Structures and Machines* in the framework of the *COMET K2 Austrian Center of Competence in Mechatronics (ACCM)* is gratefully acknowledged. The ACCM is a Research and Innovation Centre for collaborative research between science and industry in the field of mechatronics at an international level with the goal of achieving scientific knowledge and results. ACCM executes an ambitious research program in collaboration with researchers, scientific partners and companies worldwide, and served as steering organisation of the workshop. ACCM is jointly run by the following organisations:

- *Johannes Kepler University of Linz*
- *Linz Center of Mechatronics GmbH*
- *vatron gmbh*

The help of these organisations is highly appreciated. In particular, the editors wish to thank the Johannes Kepler University for serving as host of the workshop. The workshop had not been possible without the sponsoring of:

- *City of Linz*
- *Province of Upper Austria*
- *KEBA AG*
- *Magna Powertrain, Engineering Center Steyr GmbH & Co KG*
- *Salvagnini Maschinenbau GmbH*
- *SpringerWienNewYork*

Contents

Development of Integral Molding of Functionally-Graded Syntactic Foams

Tadaharu Adachi and Masahiro Higuchi

Abstract In this paper, we suggested a novel process to fabricate bulk functionally-graded (FG) syntactic plastic foams and evaluated mechanical properties of the fabricated FG foams. The density distribution in the foams was graded due to floating phenomena of the micro-balloons in the matrix resin before gelling in the fabrication process. The distribution of the density could be predicted by finite difference analysis with Richardson and Zaki's formula for Stoke's velocity. The density distribution was found to be controlled by average density, micro-balloons size, and temperature and duration of the fabrication process and so on. The progressive collapse of the FG foam due to grading mechanical properties was confirmed to be effective to improve energy absorption in the compression test.

1 Introduction

Functionally-graded materials (FGMs) [1, 2] are composite materials having continuous property distributions of two or more constituent phases. The FGMs can be designed for specific functions and applications by controlling distributions of the constituent phases. The FGMs having optimized properties can be fabricated for required condition. Then, several methods for the FGMs have been developed to fabricate appropriate distribution of properties until now [3]; electrophoretic deposition [4], chemical vapor deposition [5], sparks plasma sintering [6], centrifugal casting [7] and so on. These techniques except for the centrifugal casting would

T. Adachi (✉) · M. Higuchi
Department of Mechanical Engineering, Toyohashi University of Technology,
1-1 Hibarigaoka, Tempaku, Toyohashi 441-8580, Japan
e-mail: adachi@me.tut.ac.jp; higuchi@me.tut.ac.jp

H. Irschik et al. (eds.), *Advanced Dynamics and Model-Based Control of Structures and Machines*, DOI 10.1007/978-3-7091-0797-3_1,

not be adopted difficulty for manufacturing bulk FGMs, when thicker FGMs are in need for machine components having capability of mechanical loading. Especially, the fabrication process for the bulk FGMs must be inexpensive and simple because of requirement of large-scale production system for the machine components.

In this paper, a novel process was developed to fabricate functionally-graded (FG) syntactic plastic foams having high absorption energy, and the mechanical properties of the fabricated FG foams were measured to evaluate the fabricated foams. The foams were hollow particles (micro-balloons) filled epoxy composites, namely syntactic foams. Distributions of the micro-balloons graded due to floating the micro-balloons in low-viscous matrix resin before curing the matrix resin. In the suggested fabrication process, the distribution of the micro-balloons was graded, and the matrix resin was cured continuously after grading distribution of the micro-balloons. Effects of time and temperature in the grading process were clarified to control the distribution of the micro-balloons. Numerical analysis for distribution of the micro-balloons in the FG foams was conducted by finite difference method with Richardson and Zaki's formula [8] for Stoke's velocity. Finally, the compressive tests of the FG foams were performed to confirm their mechanical properties.

2 Fabrication concept of FG foam

The fabrication process is composed of two procedures; grating distributions of the micro-balloons and curing the matrix resin as shown in Fig. 1. After mixing the micro-balloons and the matrix resin, the mixture is heated. Increasing mobility of the micro-balloons due to decreasing viscosity of the heated matrix resin induces floating the micro-balloons in the matrix resin. As the result, distribution of the micro-balloons, namely density of the foam is graded until gelling the matrix resin. Therefore, the distribution of the micro-balloons can be controlled due to temperature dependency on the viscosity of the matrix resin before gelling. After the gelation of the matrix resin, the distribution is not changed.

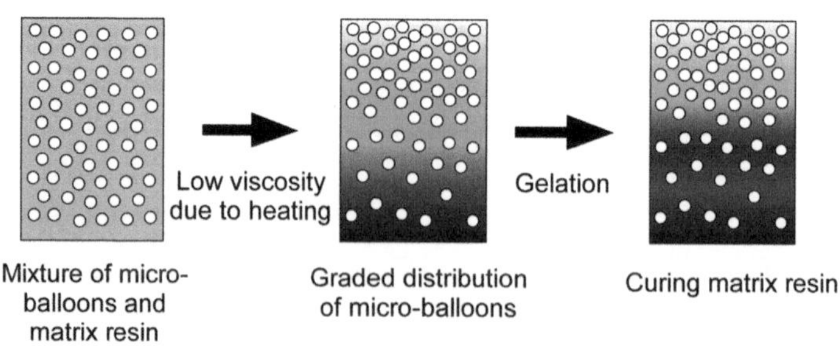

Fig. 1 Principle of fabricating FG foam

3 Fabrication of FG foam

In the experiment, several syntactic FG foams were produced by adding acrylonitrile copolymer micro-balloons coated by calcium carbonate powder (MFL-100CA, Matsumoto Yushi-Seiyaku) into epoxy resin. The diameters of the micro-balloons were 80 μm and 110 μm. The matrix epoxy resin was a bisphenol A-type epoxide resin (AER2603, Asahi Kasei e-materials) with mixture of 4-methyl hexahydrophthalic anhydride (4-methyl HHPA) and hexahydrophthalic anhydride (HHPA) (RIKACID MH-700, New Japan Chemical) as the curing agent, and 1-benzil-2-methylimidazole (BM12, Japan Epoxy Resin) as the accelerator. The weight ratio of the resin, agent, and accelerator was determined as 100:86:1 according to stoichiometry. Temperature dependence of the viscosity of the epoxy resin was measured by a rotational viscometer (DV-1+, Brookfield). The result is shown in Fig. 2. The viscosity decreased drastically from room temperature to 330 K. After that, the viscosity approached to minimum value and was kept constant. After that, gelling of the resin occurred at 370 K.

After mixing the epoxy resin and the micro-balloons sufficiently until distribution of the micro-balloons was uniform, the mixture was stored in a vacuum vessel to remove voids and was poured into an aluminum mold coated with a Teflon sheet, which was set up in an oven. The mold was 260 mm long, 5 mm wide and 130 mm deep. The weight fractions of the micro-balloons were within 0.05–0.07.

The process was done in three steps (Fig. 3). First, the mixture was kept at T_D K for t_D hours to grade the distribution of the micro-balloons. In the experiment, T_D was 338 or 323 K, and t_D was ranged from 2 h to 24 h. The process after first step was general for curing the matrix resin. The second step, pre-curing was done at 373 K for 2 h for gelling the resin, and the third step, post-curing, which greatly affects the cross-linking reaction of the resin, was done at 403 K for 15 h. Small pieces were cut from cured FG plates to measure density distribution by using Archimedian method.

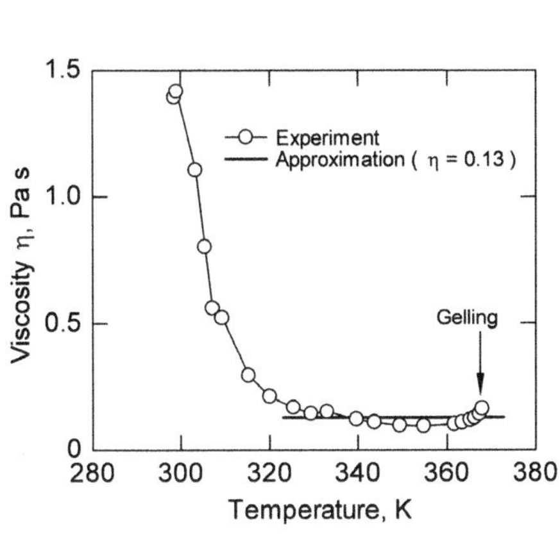

Fig. 2 Viscosity of epoxy resin before curing

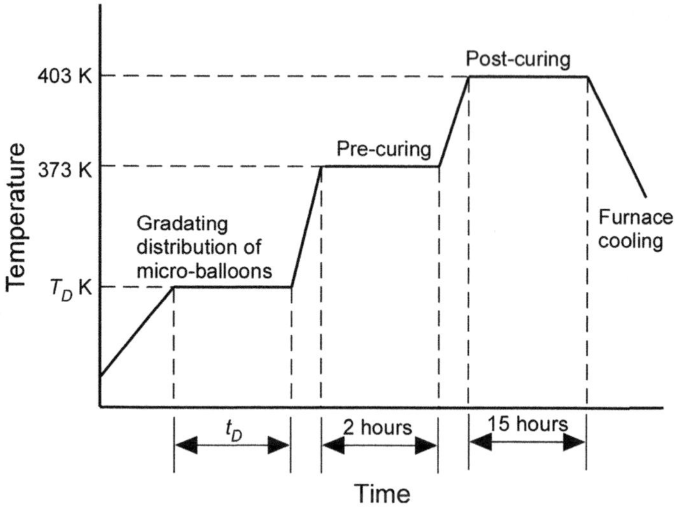

Fig. 3 Fabrication process

4 Analysis of grading density distribution

The FG foam was fabricated by utilizing floating phenomena of the micro-balloons in the matrix before gelling. Generally, floating velocity of one micro-balloon can be evaluated by the following Stoke's equations, u_s,

$$u_s = \frac{(\rho_M - \rho_B)gD^2}{18\eta}, \tag{1}$$

where η, D and ρ are the viscosity of the matrix, diameter of the micro-balloon, density, respectively. The subscript M and B denote the density of the matrix resin and the micro-balloon.

When the micro-balloons are filled into the matrix resin with high volume faction, the Stoke's velocity must be modified to evaluate average floating velocity of the micro-balloons taking interaction of the micro-balloons in the matrix resin into consideration. The modification is known as Richardson and Zaki's formula [8];

$$u = (1 - \phi)^n u_s, \tag{2}$$

where u and ϕ are the average floating velocity and the volume fraction of the micro-balloons. n is exponent dependent on Reynold's number, Re [9]. For low Reynold's number; $Re < 0.2$ or $Re < 2.0$, $n = 4.65$ is recommended generally. Then, this value of the exponent was selected in the paper.

Assuming that floating the micro-balloons is one-dimensional, the equation of continuity must be satisfied,

$$\frac{\partial \phi}{\partial t} = \frac{\partial (\phi u)}{\partial x}, \tag{3}$$

where x is vertical coordinate.

Equations (1)–(3) were calculated by finite difference method. The volume fraction ϕ of the micro-balloons was discretized by $x = i\Delta x$ and $t = j\Delta t$ $(i, j = 0, 1, 2, ...)$

$$\phi(i\Delta x, j\Delta t) = \phi_i^j. \tag{4}$$

By substituting Eq. (4) into Eqs. (2) and (3), the difference equation are obtained as follows;

$$\phi_i^{j+1} = \left\{1 + \left(1 - \phi_i^j\right)^n \lambda\right\} \phi_i^j - \left(1 - \phi_{i-1}^j\right)^n \phi_{i-1}^j \lambda, \tag{5}$$

where $\lambda = u_s \Delta t / \Delta x$.

In the numerical analysis, the density of the micro-balloons and the epoxy resin were 130 and 1,190 kg/m^3, respectively. The viscosity of the matrix epoxy resin was assumed to be constant; $\eta = 0.13$ Pas (Fig. 2). In the numerical analysis, the depth of the FG foam was 130 mm and divided into 600 nodes. The increment time, Δt was 3 s. Because the initial distribution of the micro-balloons was uniform, the initial condition was given as

$$\phi_i^0 = \phi^0, \tag{6}$$

where ϕ^0 is average volume fraction.

Because the bottom of the mold was the origin of x, the boundary condition was

$$\phi_0^j = 0. \tag{7}$$

In the analysis, the maximum value of the volume fraction was assumed to be 0.632 which was determined by the experiment.

5 Density distribution

The density distributions of the foams fabricated under different conditions are shown with the numerical results in Figs. 4–6. The grading distributions of the density were confirmed to be yielded in the foam. The different distributions were also confirmed to be obtained due to different conditions; average density and fabrication process. When the average density was low, the distribution varied rapidly according to Eq. (2). Then, the distribution in Fig. 5 was gradual compared with the one in Fig. 4. If the micro-balloons diameter was larger, the distribution would be also gradual. After long time passed in the first step of the process, the density distribution in the foams was divided into two phases as shown in Fig. 6.

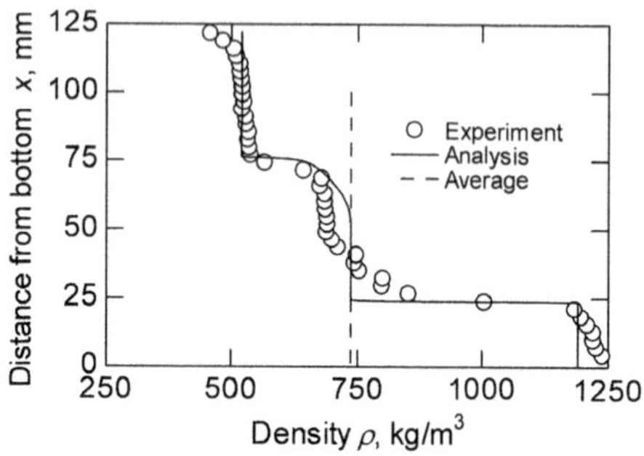

Fig. 4 Density distribution. Process: $T_D = 338$K, $t_D = 2$h, average density 736 kg/m^3

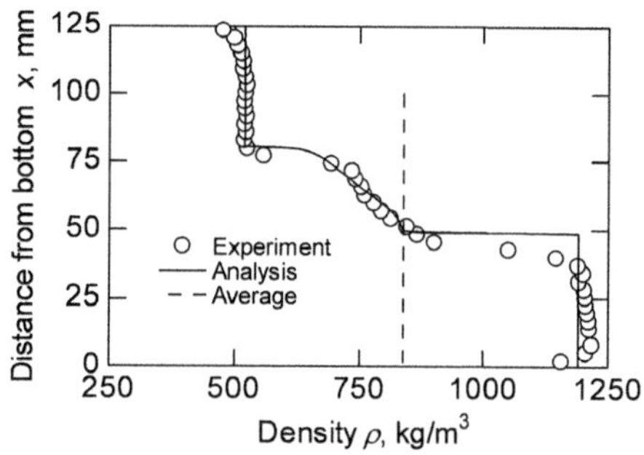

Fig. 5 Density distribution. Process: $T_D = 338$K, $t_D = 2$h, average density 838 kg/m^3

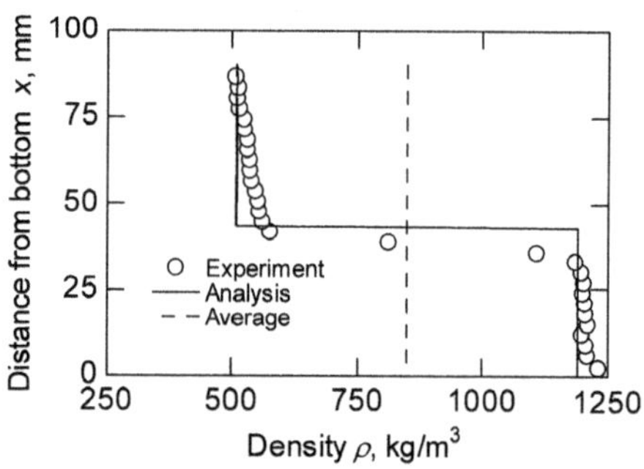

Fig. 6 Density distribution. Process: $T_D = 323$K, $t_D = 12$h, average density 848 kg/m^3

The numerical results in Figs. 4–6 agreed with the experimental results. By using calculation based on Richardson and Zaki's formula, the distribution could be predicted generally. Therefore, the distribution of the mechanical properties in the foams would be able to be calculated from the density distribution, since mechanical properties of foams; elastic modulus, strength, and so on, can be evaluated by approximate equation [10].

6 Compression tests

Compression test of the fabricated FG foams were conducted to confirm availability of the foams. Two specimens were cut from the fabricated plate. One specimen was FG foam (Fig. 7) and the other was uniform (Fig. 8). Since the FG foam collapsed progressively from low density region to high density one due to compression, the compressive stress gradually increased after yielding. Then the progressive collapse caused high energy absorption. On the other hand, the whole specimen having uniform distribution of the density collapsed and the plateau region occurred on the stress–strain curve after yielding. Therefore, the FG foams were found to be effective for materials of mechanical components because the FG foams had high energy absorption under compression.

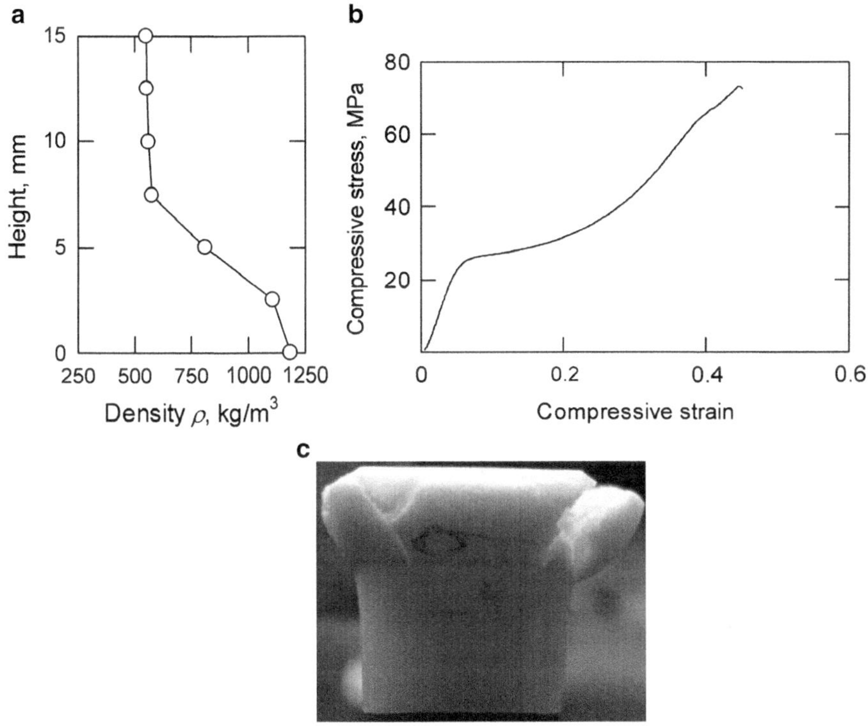

Fig. 7 Compression test of FG foam: (**a**) density distribution; (**b**) compressive stress–strain curve; (**c**) specimen after test

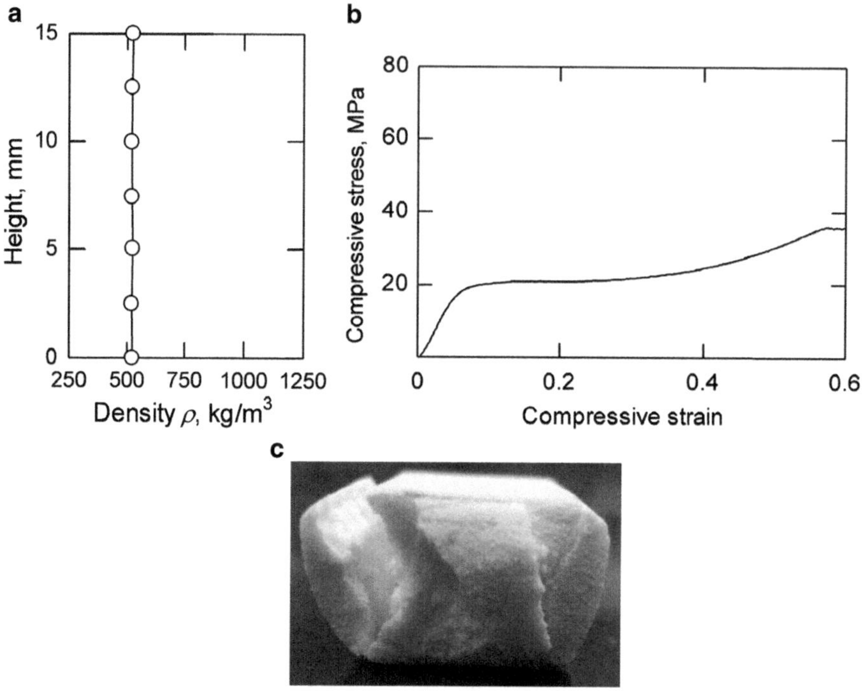

Fig. 8 Compression test of uniform foam: (**a**) density distribution; (**b**) compressive stress–strain curve; (**c**) specimen after test

7 Conclusion

In this paper, we suggested a novel process to fabricate bulk FG syntactic plastic foams, and evaluated the mechanical properties of the fabricated FG foams. The density distribution in the foams was graded by floating phenomena of the micro-balloons before gelling the matrix resin. The distribution of the micro-balloons could be predicted by finite difference analysis with Richardson-Zaki's formula for Stoke's velocity. The progressive collapse of the FG foam due to grading mechanical properties was confirmed to be effective to improve energy absorption.

References

1. Markworth, A.J., Ramesh, K.S., Parks, W.P.: Review: modeling studies applied to functionally graded materials. J. Mater. Sci. **30**, 2183–2193 (1995)
2. Birman, V., Byrd, L.W.: Modeling and analysis of functionally graded materials and structures. Appl. Mech. Rev. **60**, 195–216 (2007)

3. Kieback, B., Neubrand, A., Riedel, H.: Processing techniques for functionally graded materials. Mater. Sci. Eng. A, **362**, 81–105 (2003)
4. Put, S., Vleugels, J., Van der Biest, O.: Microstructural engineering of functionally graded materials by electrophoretic deposition. J. Mater. Process. Technol. **143–144**, 572–577 (2003)
5. Kim, J.I., Kim, W.J., Choi, D.J., Park, J.Y., Ryu, W.S.: Design of a C/SiC functionally graded coating for the oxidation protection of C/C composites. Carbon **43**, 749–1757 (2005)
6. Shen, Z.J., Nygren, M.: Laminated and functionally graded materials prepared by spark plasma sintering. Key Eng. Mater. **206**, 2155–2158 (2002)
7. Biesheuvel, P.M., Verweij, H.: Calculation of the composition profile of a functionally graded material produced by centrifugal casting. J. Am. Ceram. Soc. **83**, 743–749 (2000)
8. Richardson, J.F., Zaki, W.N.: Sedimentation and fluidization: Part I. Trans. Instrum. Chem. Eng. **32**, 35–53 (1954)
9. Siwiec, T.: The experimental verification of Richardson-Zaki Law on example of selected beds used in water treatment. Elec. J. Polish Agric. Univ. **10**(2), #5 (2007)
10. Gibson, L.J., Ashby, M.F.: Cellular Solids, 2nd ed. Cambridge University Press, Cambridge (1999)

Dynamic Instabilities of Simple Inelastic Structures Subjected to Earthquake Excitation

Christoph Adam and Clemens Jäger

Abstract The seismic collapse capacity of highly inelastic non-deteriorating single-degree-of-freedom (SDOF) systems, which are vulnerable to the destabilizing effect of gravity loads (P-delta effect), is evaluated. The collapse capacity based on different definitions of the intensity measure is discussed. In particular, collapse capacity spectra are derived utilizing sets of Incremental Dynamic Analyses (IDAs) involving 44 recorded ground motions.

1 Introduction

A main objective of earthquake engineering is to provide the structure with an adequate margin of safety against collapse [1]. Observations of collapsed buildings in severe earthquake events reveal that sidesway collapse is the main mode of structural collapse. Sidesway collapse can be a result of successive reduction of the load bearing capacity of structural components. In inelastic flexible structures gravity loads acting through lateral displacements amplify structural deformations and stress resultants (P-delta effect), and thus may be another source of sidesway collapse. In many realistic cases collapse is a consequence of combined action of both effects [1].

In this study the focus is on collapse of highly inelastic single-degree-of-freedom (SDOF) structures, which are vulnerable to the P-delta effect. A profound insight into the P-delta effect on the inelastic seismic response of structures is given e.g. by Bernal [2], and Ibarra and Krawinkler [3]. Asimakopoulos et al. [4] and Villaverde [5] provide an overview on studies dealing with collapse by dynamic instability

C. Adam · C. Jäger
Unit of Applied Mechanics, University of Innsbruck, 6020 Innsbruck, Austria
e-mail: christoph.adam@uibk.ac.at; clemens.jaeger@uibk.ac.at

H. Irschik et al. (eds.), *Advanced Dynamics and Model-Based Control of Structures and Machines*, DOI 10.1007/978-3-7091-0797-3_2,

in earthquake excited structures. Fundamental studies of the effect of gravity on inelastic SDOF systems subjected to earthquakes have been presented in Bernal [6] and MacRae [7]. Miranda and Akkar [8] present an empirical equation to estimate the minimum lateral strength up to which P-delta induced collapse of SDOF systems is prevented. In Adam et al. [9, 10] so-called collapse capacity spectra have been introduced for the assessment of the seismic collapse capacity of SDOF structures.

In the present paper emphasis is given on the collapse capacity. Collapse capacity spectra based on different definitions of the intensity measure are derived utilizing a set of 44 ordinary ground motion records. Results and conclusions of this study are valid only for non-deteriorating cyclic behavior, i.e. strength and stiffness degradation is not considered.

2 Framework and definitions

2.1 The P-delta effect in inelastic SDOF systems

In the following the effect of a gravity load on an SDOF oscillator with inelastic spring characteristics is demonstrated considering an inverted mathematical pendulum of length h as shown in Fig. 1a [9, 10]. An inelastic rotational spring with initial stiffness k_r supports the bottom of the rigid rod of the pendulum, and a rotational damper with parameter r_r models structural viscous damping. For small angles, $|\varphi| \ll 1$, the horizontal displacement x of the lumped mass m serves as the characteristic quantity of deformation. The gravity load P generates a shearing of the hysteretic force-displacement $(f - x)$ relationship. Characteristic displacements (such as the yield displacement x_y) of this relationship remain unchanged, whereas

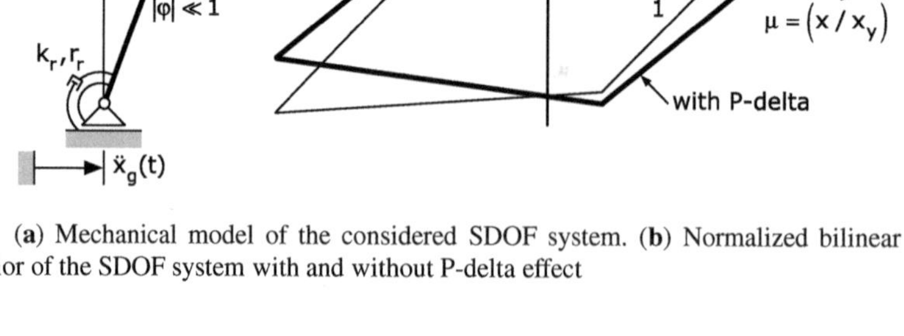

Fig. 1 (**a**) Mechanical model of the considered SDOF system. (**b**) Normalized bilinear cyclic behavior of the SDOF system with and without P-delta effect

the characteristic forces (such as the strength f_y) are reduced. The slope of the curve is decreased in its elastic and post-elastic branch of deformation. The magnitude of this reduction can be expressed by the stability coefficient [7, 10]:

$$\theta = \frac{Ph}{k_r} \tag{1}$$

As a showcase Fig. 1b visualizes the P-delta effect on the hysteretic behavior of an SDOF system with non-deteriorating bilinear characteristics. In this example the post-yield stiffness is negative because the stability coefficient θ is larger than the hardening ratio α.

A negative slope of the post-tangential stiffness, expressed by the difference of the stability coefficient θ and the strength hardening coefficient α, $\theta - \alpha$, is the essential condition that the structure may collapse under severe earthquake excitation. In [10] it is shown that collapse of inelastic SDOF systems vulnerable to P-delta is mainly governed by the following parameters:

- The negative slope of the post-tangential stiffness $\theta - \alpha$,
- the elastic structural period of vibration T,
- the viscous damping coefficient ζ (usually taken as 5%), and
- the shape of the hysteretic loop.

2.2 *Intensity measure and collapse capacity*

There is no unique definition of intensity of an earthquake record. Examples of the intensity measure are the 5% damped spectral acceleration S_a at the structure's period T, S_a $(T, \zeta = 0.05)$, the peak ground acceleration (PGA), the peak ground velocity (PGV), and the peak ground displacement (PGD), see e.g. [11]. An appropriate choice of the intensity measure exhibits a narrow distribution of deformation demands induced by a set of several earthquake records.

The collapse capacity is defined as the maximum ground motion intensity at which the structure still maintains dynamic stability [1]. The Incremental Dynamic Analysis (IDA) procedure [12] is applied to predict the collapse capacity. In an IDA for a given structure and a given acceleration time history of an earthquake record dynamic time history analyses are performed repeatedly, where in each subsequent run the intensity of the ground motion is incremented. As an outcome a characteristic intensity measure is plotted against the corresponding maximum characteristic structural response quantity for each analysis. The procedure is stopped, when the response grows to infinity, i.e. structural failure occurs. The corresponding intensity measure is referred to as collapse capacity of the considered structure for this specific ground motion record (denoted by i). In non-dimensional form the ith collapse capacity may be defined according to

$$CC|_{Sa(T),i} = \left.\frac{S_{a,i}(T)}{g\gamma}\right|_{collapse}, \quad CC|_{PGA,i} = \left.\frac{PGA_i}{g\gamma}\right|_{collapse}$$

$$CC|_{PGV,i} = \left.\frac{\omega\, PGV_i}{g\gamma}\right|_{collapse}, \quad CC|_{PGD,i} = \left.\frac{\omega^2\, PGD_i}{g\gamma}\right|_{collapse} \tag{2}$$

γ represents the yield strength coefficient, $\gamma = f_y/(mg)$, g is the acceleration of gravity, and ω denotes the structural circular frequency, $\omega = 2\pi/T$.

The inherent record-to-record variability leads to different collapse capacities for different ground motion records. Thus, the collapse capacities are determined for an entire set of n ground motion records, and subsequently evaluated statistically. For example, the median of the individual collapse capacities CC_i may be considered as representative collapse capacity CC for the analyzed structure and set of ground motion records. About 16% and 84% percentiles characterize the distribution of the individual collapse capacities.

As an example Fig. 2a shows IDA curves of an SDOF structure with bilinear hysteretic loop and the following structural parameters: $T = 1.0s$, $\zeta = 0.05$, $\theta - \alpha = 0.10$. In this example the intensity measure is the 5% damped spectral acceleration S_a at T. All results of this study are based on the ATC63 far-field (ATC63-FF) set of ordinary ground motions [13]. The records of this set originate from severe seismic events of moment magnitude between 6.5 and 7.6 and closest distance to the fault rupture larger than 10 km. Thereby, only strike-slip and reverse sources are considered. All 44 records of this set were recorded on NEHRP site classes C (soft rock) and D (stiff soil).

Different characteristics of the individual ground motion records lead to different IDA curves, which are plotted in grey, see Fig. 2a. A horizontal tangent of an IDA curve indicates structural failure. In this example the individual collapse capacities vary between 1.8 and 6.5. Furthermore, the median, 16% and 84% percentile curves are shown in black. Here, the intensity measure, where the median IDA curve becomes horizontally, is referred to as characteristic collapse capacity: $CC|_{Sa(T)} = 2.8$.

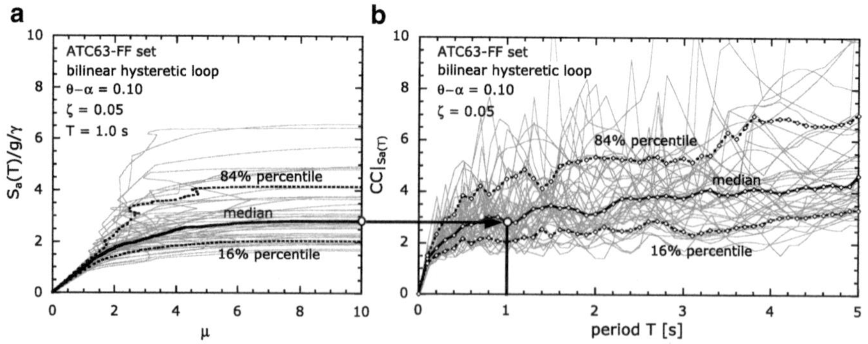

Fig. 2 (**a**) IDA curves, and (**b**) collapse capacity spectra. SDOF system, ATC63-FF set of ground motion records. Individual outcomes for each record and statistically evaluated results

3 Collapse capacity spectra

The representation of the collapse capacity of SDOF systems with assigned ζ and $\theta - \alpha$, and a particular hysteretic loop as a function of the initial period T results in collapse capacity spectra. Adam et al. [9, 10] introduce the concept of collapse capacity spectra for the assessment of the collapse capacity of non-deteriorating SDOF systems vulnerable to the P-delta effect. Exemplarily, Fig. 2b shows collapse capacity spectra of a bilinear SDOF system ($\zeta = 0.05$, $\theta - \alpha = 0.10$) for the 44 ground motions of the ATC63-FF set. Additionally, median, 16% and 84% percentile spectra are displayed in the period range between 0 and $5s$.

Subsequently, collapse capacity spectra according to the definitions of (2) are compared and evaluated. Figure 3 reveals the different characteristics of median collapse capacity spectra exemplarily for SDOF systems with a negative post-yield slope of $\theta - \alpha = 0.06$ and a damping coefficient of $\zeta = 0.05$. Collapse capacities, which rely on PGA as intensity measure show a steep rise with increasing period. On the other hand, the graph of the median collapse capacity spectra based on S_a $(T, \zeta = 0.05)$ exhibits a decreasing gradient with increasing period. An appropriate intensity measure for a carefully selected set of ground motions leads to a narrow distribution of the individual collapse capacities. Figure 4 shows the characteristic quantity s of the distribution for the collapse capacity spectra presented in Fig. 3. Here, s is defined as the square root of the ratio of 84% percentile to 16% percentile collapse capacities,

$$s = \sqrt{\frac{CC|_{84\%}}{CC|_{16\%}}} \tag{3}$$

The larger the deviation of s from 1 the larger is the scatter of the individual results. From Fig. 4 it can be seen that s is for the collapse capacities based on the spectral

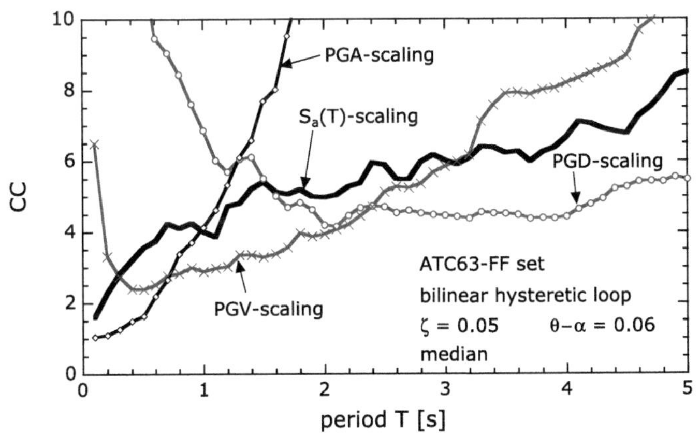

Fig. 3 Median collapse capacity spectra based on different definitions of the intensity measure

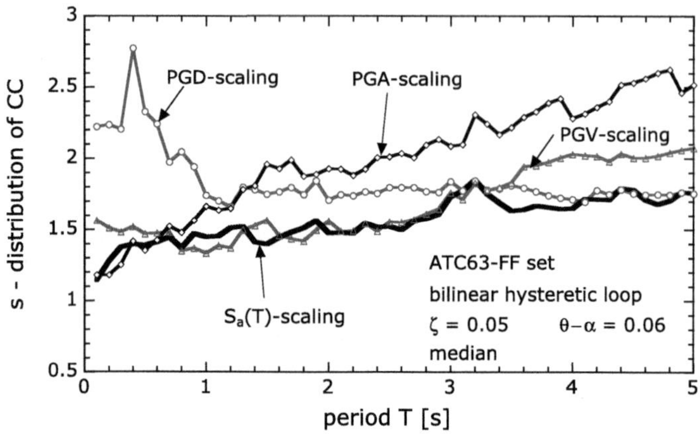

Fig. 4 Distribution of collapse capacity spectra based on different definitions of the intensity measure

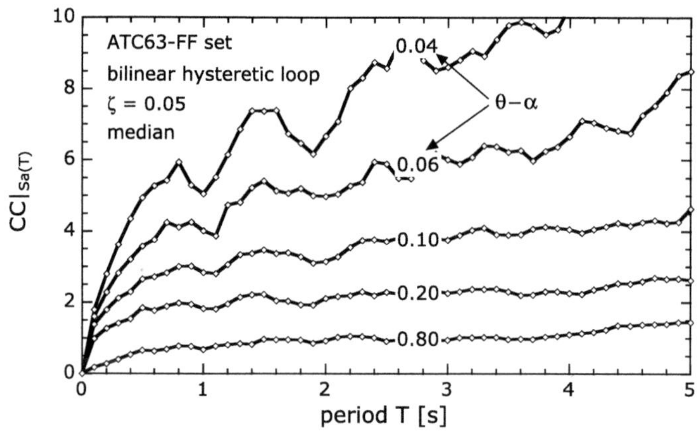

Fig. 5 Collapse capacity spectra based on the 5% damped spectral acceleration S_a at the structure's period T

acceleration a minimum in the nearly entire period range. For very stiff systems s is about 1.2 and then increases steadily with increasing period up to 1.75 at $T = 5.0s$. From this result it can be concluded that this definition of intensity measure is favorable for predicting the collapse capacity of SDOF structures vulnerable to P-delta. The other definitions of the intensity measures lead only at certain period segments to distributions comparable with S_a $(T, \zeta = 0.05)$. In the acceleration sensitive small period range PGA is an appropriate intensity measure, PGV in the velocity sensitive period range between 0.8 and 3.2s, and PGD leads in the large period range to a distribution similar to S_a $(T, \zeta = 0.05)$.

Figures 5–7 show collapse capacity spectra for a series of negative post-yield slopes $\theta - \alpha$ ranging from 0.04 to 0.80. The collapse capacities of these Figures

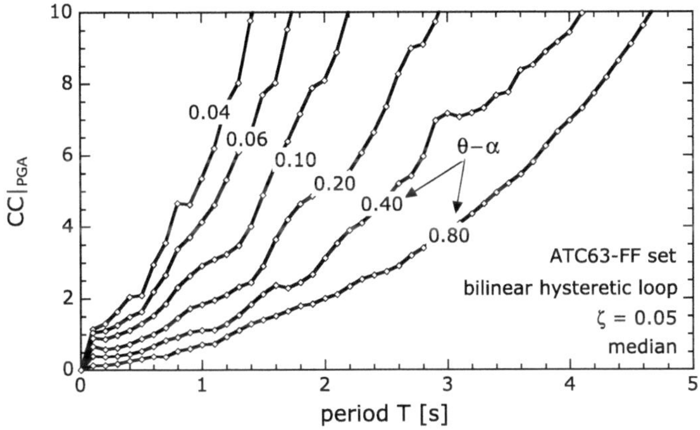

Fig. 6 Collapse capacity spectra based on the peak ground acceleration (PGA)

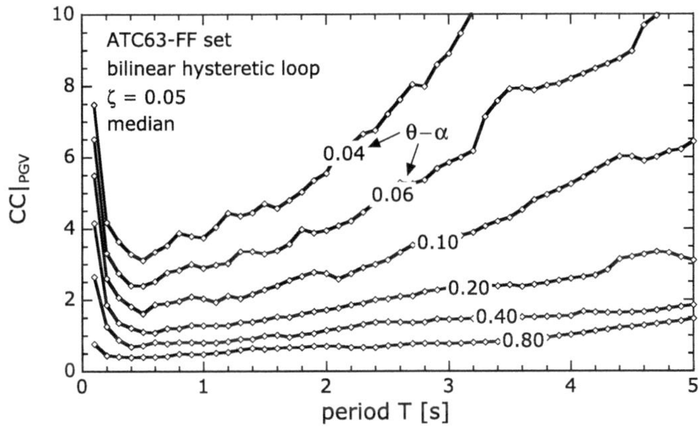

Fig. 7 Collapse capacity spectra based on the peak ground velocity (PGV)

rely on different definitions of the intensity measure: Fig. 5 is based on the 5% damped spectral acceleration S_a at the fundamental period T, Fig. 6 on PGA, and Fig. 7 on PGV. All representations have in common that an increasing slope of the negative stiffness reduces the collapse capacity. However, the run of the graphs is different for each definition, compare with Figs. 5–7. In particular, rigid systems ($T = 0$) exhibit for acceleration dependent intensity measures (i.e. $S_a(T)$ and PGA) a collapse capacity of $CC = 0$, whereas the application of PGV leads to a collapse capacity of infinity. This can be attributed to the fact that the definition of the PGV based collapse capacity is multiplied by the structure's fundamental frequency ω, which is infinity for $T = 0$, compare with (2).

Regression analyses convert collapse capacity spectra in design collapse capacity spectra with "smooth curves", compare with [10]. Application of design collapse capacity spectra is simple: an estimate of the elastic period of vibration T, stability coefficient θ and hardening ratio α of the actual SDOF structure need to be determined. Subsequently, from the chart the corresponding collapse capacity CC can be read [10].

In [14] it is shown that collapse capacity spectra can be applied to assess the collapse capacity of multi-degree-of-freedom (MDOF) systems vulnerable to P-delta, provided that a corresponding equivalent SDOF system reflects with sufficient accuracy the dynamic behavior of the MDOF system. In general, regular structures satisfy this requirement since the global P-delta effect is mainly governed by the fundamental mode. For further details see [14].

References

1. Krawinkler, H., Zareian, F., Lignos, D.G., Ibarra, L.F.: Prediction of collapse of structures under earthquake excitations. In: Papadrakakis M., Lagaros, N.D., Fragiadakis, M. (eds.) Proceedings of the 2nd International Conference on Computational Methods in Structural Dynamics and Earthquake Engineering (COMPDYN 2009), Rhodes, Greece, CD-ROM paper, paper no. CD449, 22–24 June 2009
2. Bernal, D.: Instability of buildings during seismic response. Eng. Struct. **20**, 496–502 (1998)
3. Ibarra, L.F., Krawinkler, H.: Global collapse of frame structures under seismic excitations. Report No. PEER 2005/06, Pacific Earthquake Engineering Research Center, University of California, Berkeley, CA, 2005
4. Asimakopoulos, A.V., Karabalis, D.L., Beskos, D.E.: Inclusion of the P-Δ effect in displacement-based seismic design of steel moment resisting frames. Earthquake Eng. Struct. Dyn. **36**, 2171–2188 (2007)
5. Villaverde, R.: Methods to assess the seismic collapse capacity of building structures: State of the art. J. Struct. Eng. **133**, 57–66 (2007)
6. Bernal, D.: Amplification factors for inelastic dynamic P-Δ effects in earthquake analysis. Earthquake Eng. Struct. Dyn. **15**, 635–651 (1987)
7. MacRae, G.A.: P-Δ effects on single-degree-of-freedom structures in earthquakes. Earthquake Spectra, **10**, 539–568 (1994)
8. Miranda, E., Akkar, S.D.: Dynamic instability of simple structural systems. J. Struct. Eng. **129**, 1722–1726 (2003)
9. Adam, C.: Global collapse capacity of earthquake excited multi-degree-of-freedom frame structures vulnerable to P-delta effects. In: Yang, Y.B. (ed.) Proceedings of the Taiwan – Austria Joint Workshop on Computational Mechanics of Materials and Structures, National Taiwan University, Taipei, Taiwan, pp. 10–13. 15–17 November 2008
10. Adam, C., Jäger, C.: Seismic collapse capacity of basic inelastic structures vulnerable to the P-delta effect, Earthquake Eng. Struct. Dyn. (accepted for publication)
11. Yakut, A., Yilmaz, H.: Correlation of deformation demands with ground motion intensity. J. Struct. Eng. **134**, 1818–1828 (2008)
12. Vamvatsikos, D., Cornell, C.A.: Incremental dynamic analysis. Earthquake Eng. Struct. Dyn. **31**, 491–514 (2002)
13. ATC63: Quantification of Building Seismic Performance Factors. Applied Technology Council ATC 63 Project Report – 90% Draft. FEMA P695, April 2008
14. Adam, C., Jäger, C.: Seismic induced global collapse of non-deteriorating frame structures. In: Papadrakakis, M., Lagaros, N.D., Fragiadakis, M. (eds.) Computational Methods in Earthquake Engineering, vol. 99, pp. 21–40. Springer, Dordrecht (2011)

Nonlinear Dynamics and Control of Ultrasonic Technology Processes and Systems

V.K. Astashev

Abstract Some problems of the dynamics and control of ultrasonic processing systems are discussed. Ultrasonic systems as special kinds of vibratory machines working under nonlinear technology load are considered. The influence of ultrasonic vibration on working processes is investigated using rheological models, reflecting the materials real elastic and plastic properties. The theoretical results are compared with experimental data. Amplitude-frequency characteristics of ultrasonic systems and specific nonlinear dynamical effects are discussed. The possibilities of the ultrasonic machines autoresonant excitation using a feedback system are considered.

1 Ultrasonic technology processes and devices

The working process of an ultrasonic machine is performed by subjecting its tool to a combination of two motions. A driving motion program is required to shape the workpiece. A high-frequency (ultrasonic) vibration of specific direction, frequency and intensity is then superimposed. These methods are used in the erosion treatment of brittle materials [13], processes plastic deformation [14], metal cutting operation [11], wire and tube drawing [12] etc.

The construction of the machine and its elements depends critically on the process being performed by the tool. Figure 1a, b shows the arrangement of ultrasonic machines used for plastic deformation and turning hard-to-machine materials. A piezoelectric transducer 3 generates vibration which is then transmitted to the tool 4 via a waveguide-concentrator 2 with an increase in amplitude. The whole oscillating system is fastened in the body 5 of the acoustic head. During

V.K. Astashev (✉)
Institute of Machines Science of RAS, Maly Khariton'evsky per. 4, Moscow, Russia
e-mail: v_astashev@mail.ru

H. Irschik et al. (eds.), *Advanced Dynamics and Model-Based Control of Structures and Machines*, DOI 10.1007/978-3-7091-0797-3_3,

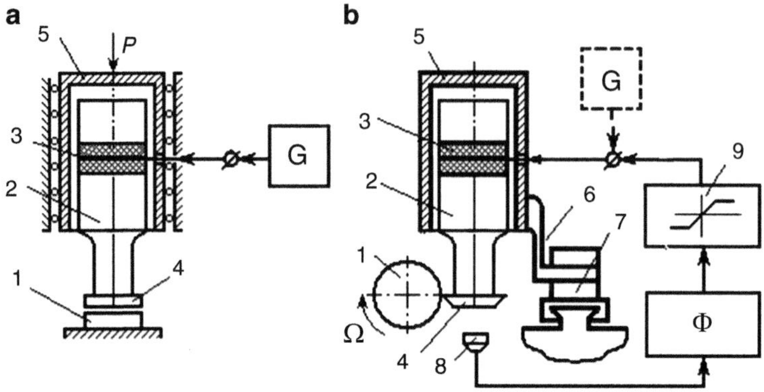

Fig. 1 Arrangement of ultrasonic machines for plastic deformation (a) and turning

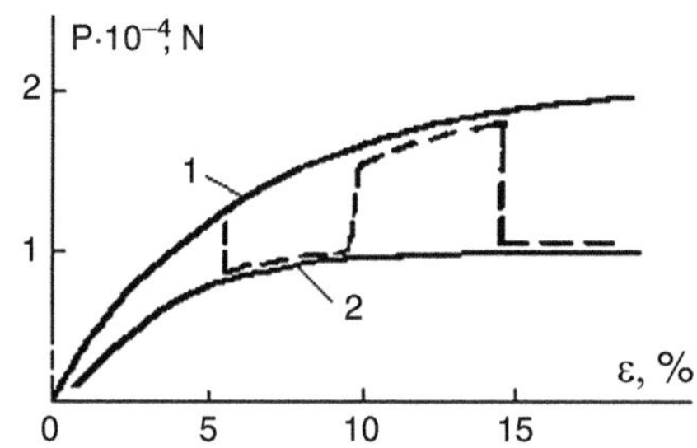

Fig. 2 Strain-force diagrams of static (curve 1) and ultrasonic (curve 2) deformation

processing the tool moves towards the workpiece with a driving speed ν or subjects to a static driving force P. In such a way kinematic or power drive is realized.

The main peculiarity of ultrasonic technology processes consist in the decrease in static force necessary to fulfill the processes where plastic deformation of the treated medium takes place. This has been demonstrated in numerous experiments.

The results of the experiment [10] under deformation of a specimen is shown in Fig. 2.

Curve 1 shows a "strain–force" diagram, obtained for a static (with no vibration) compression of an aluminium specimen of diameter $d = 14$ mm and height $H = 20$ mm. The relative strain $e = h/H$ is shown on the x-axes, where h is the shortening of the specimen. Curve 2 shows the change in static force P during compressing a similar specimen in the presence of ultrasonic vibration. In both cases the compression process was carried out with the same constant speed $\nu = 0.5$ mm/min.

Note that in experiments the vibration employed was at a frequency of $f = 22$ kHz, which is usual for an ultrasonic process. The amplitude at the working end of the waveguide was $a = 10$ μm. Vibration at such a high frequency and small amplitude are indistinguishable by an unaided eye. The observer has the impression that the softening of the material and the change of its elastic-plastic properties take place under the influence of ultrasound. It is important that such a softening

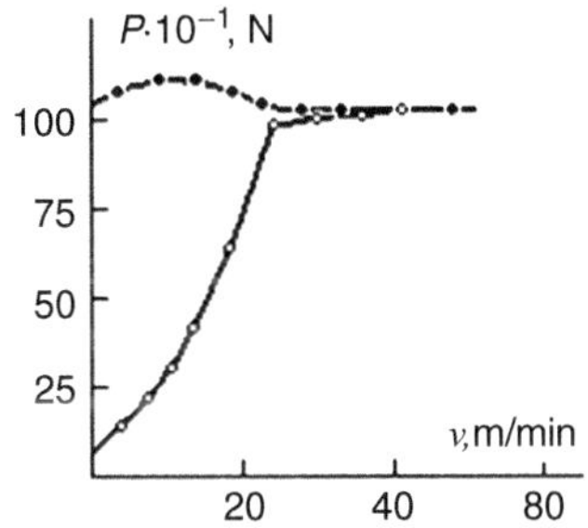

Fig. 3 The dependence of cutting force P on the cutting speed v

can only be observed in the presence of vibration. The results of an experiment investigating the shortening of specimens under periodically imposed ultrasonic vibration are shown by a dashed line and confirm the previous observation.

The described properties are typical for all processes in which the tool is excited with ultrasonic vibration. They are also present in cutting processes when ultrasonic vibration is superimposed on to the cutting tool. The dependence of cutting force P on the cutting speed v obtained in experiments [9] is shown in Fig. 3.

Curve 1 relates to traditional turning, while curve 2 is encountered when ultrasonic vibration is superimposed on to the cutting tool during turning. An important observation of these experiments is a "disappearance" of the cutting force for values of speed v close to zero. Here, of course, a constant component of cutting force is considered. This is either measured from the torque experienced by the workpiece, or by the strain experienced by the cutter during turning.

As the cutting speed is increased to the value $v = \alpha\omega$, where ω is the angular frequency, the cutting force monotonically increases up to the value that it would take in the absence of vibration. Such a relation between a constant force and the relative speed of displacement is typical for systems with superimposing vibration. We see, from experimental results, that the static force is almost completely eliminated when speeds v are low in the processes of plastic deformation with the application of ultrasonics.

A physical mechanism for this phenomenon can be modelled [2, 3] on basis of real elastic-plastic properties of materials.

2 Nonlinear models of ultrasonic technology processes

Let us consider the process of material deformation under vibrational loading using the characteristic of an ideal elastic-plastic material. The motion of tool with respect to the deformed specimen in a general case is described as follows

$$u(t) = vt + u^0(t) = vt + a \sin \omega t, \tag{1}$$

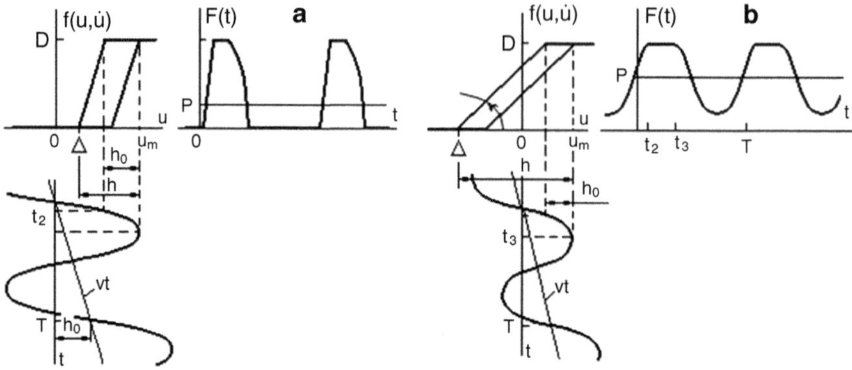

Fig. 4 Force of interaction creation during process of impulse (a) and continuous (b) deformation

where ν is a feed speed, $u^0(t)$ is a law of vibrational motion, a is the amplitude and ω is the angular frequency.

Figure 4 illustrates the possible occurrences when specimen deformation takes place using a vibrating tool. Figure 4 shows a graph of the dynamic characteristic $f = f(u, \dot{u})$, a motion law graph below it, and a time dependence the force of interaction $F(t)$ on the right-hand side. Here h is the all deformation and h_0 is the residual deformation for a period of vibration.

Due to the periodic nature of the processes under consideration, the force of interaction

$$F(t) = f[u(t), \dot{u}(t)]$$

is a periodic function of time with a period $T = 2\pi/\omega$. Taking this into account, and using the theorem of momentum, the relation between the constant force P and the parameters of motion of the tool is obtained in the following form:

$$P = \frac{1}{T}\int_0^T F(t)dt = \frac{1}{T}\int_0^T f[u(t), \dot{u}(t)]dt. \tag{2}$$

The average speed of deformation is $\nu = h_0/T = h_0\omega/2\pi$.

We can see two type of deformation regimes: the regimes of impulse (vibro-impact) (Fig. 4a) and continuous (impact-free) (Fig. 4a) deformation. The results of calculations are shown in Fig. 5, where the dependence of the static force P on plastic deformation speed ν is shown for different values of D/k_0a, where D is force of plastic deformation and k_0 is static stiffness of the specimen. The dashed line separates the regions of vibro-impact (lower region) and continuous (higher region) deformation.

It must be noted that both continuous and impulsive regimes were observed in a number of experiments [8] (Rosenberg et al. 1963). In each case it was found that impulsive regimes are more efficient. The relations found have real solutions

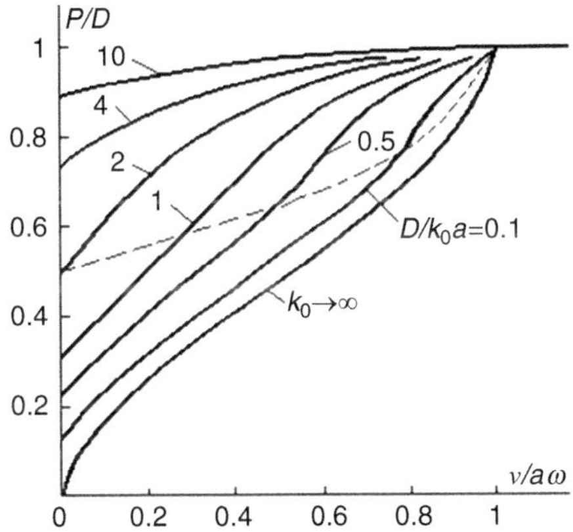

Fig. 5 The dependence of cutting force on the cutting speed in the ultrasonic turning

for deformation speeds $v < a\omega$. If $v = a\omega$ we get $P = D$ and further increase in speed v does not change the static force P required for plastic deformation as this is observed in numerous experiments.

The suggested approach can be used for the description of various ultrasonic processes like to turning and wire drawing in which both forces of deformation, and frictional forces are pronounced. The results obtained here allow us to explain many of the effects observed experimentally. As an example, the plots in Fig. 5 are close to the dependence of cutting force, on cutting speed in the turning of a workpiece with the imposition of ultrasonic vibration applied to the cutting tool.

3 Amplitude-frequency characteristics of ultrasonic systems

Due to their construction, vibration systems are characterized by a high Q-factor. Therefore, ultrasonic machines can only work effectively in resonant regimes. In practice, ultrasonic machines are adjusted to resonance mostly under idle running, assuming that the influence of the loads in operating conditions is negligible. However, as it was established [4, 8] this assumption is incorrect. It is shown that loads lead, not only to a change in the resonant frequencies of the vibrating system, but cause specific nonlinear distortions of its amplitude-frequency characteristics [1,5].

To find amplitude-frequency characteristics ultrasonic device dynamics can be considered using generalized dynamic model where all forces and dynamic stiff nesses are reduced to one element namely to the tool. Its movement of a kind (1) is described by the equation

$$u(t) = W^{-1}(j\omega)\left[N(j\omega)e^{j\omega t} - f(u, \dot{u})\right] \tag{3}$$

where $N(j\omega)e^{j\omega t}$ is exciting force complex amplitude whose effect has been relocated to the tool and $W(j\omega)$ is the vibration system relocated dynamic stiffness.

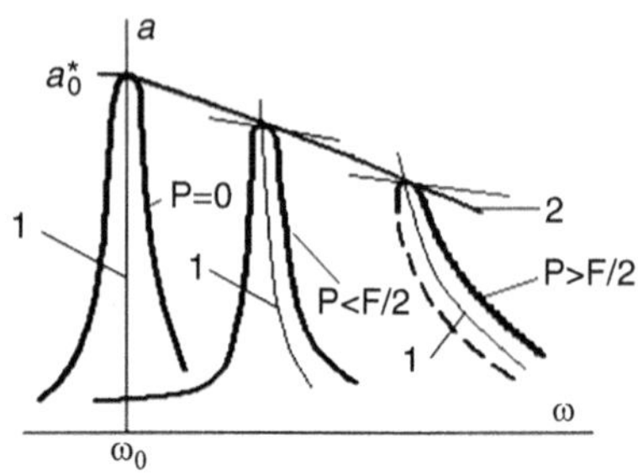

Fig. 6 The amplitude-frequency characteristics of system with power drive for different values of static force

Take account of the nature of the solutions sought the harmonic linearization of the nonlinear dynamic characteristic may be undertaken

$$f(u, \dot{u}) \approx P(v, a) + \left[k(v, a) + j\omega b(v, a)\right]u^0. \tag{4}$$

Here $P(v, a)$ is a constant component defined by (2); $k(v, a)$ and $b(v, a)$ are the equivalent elastic and dissipative components of the technology load

After substituting (4) into (3) the amplitude of the tool vibration can be founded

$$a = \left| \frac{N(j\omega)}{U(\omega) + k(v, a) + j\omega[V(\omega) + b(v, a)]} \right|, \tag{5}$$

where $U(\omega) = \text{Re}\, W(j\omega)$ and $V(\omega) = \omega^{-1}\text{Im}\, W(j\omega)$.

The amplitude-frequency characteristics system with power drive for different values of static force P are shown in Fig. 6.

When relatively small forces $P < N/2$ are acted the shapes of the resonant curves are the same under load or during idling motion ($P = 0$) but their maximum values are shifted to higher frequencies as P is increased. A further increase in force ($P > N/2$) drastically changes the resonant curves character and leads to appearances an unstable branch shown by a dashed line. Curves 1 are the backbone curves that define the dependence of natural frequency on amplitude. Line 2 is the resonant curves envelope curve.

Figure 7 shows the resonant curves transformation for ultrasonic machine with cinematic drive [6, 7], when driving speed v is increased from $v = 0$ to value that exceed the tool vibration speed in idling resonant regime ($v = a_0 w_0$).

4 Autoresonant excitation and stabilization of resonant modes

It should be noted that obtaining of stable and predictable results of the ultrasonic processing is possible only during the realization of the most effective resonant modes under conditions of a varying technological load. Difficulties of resonant

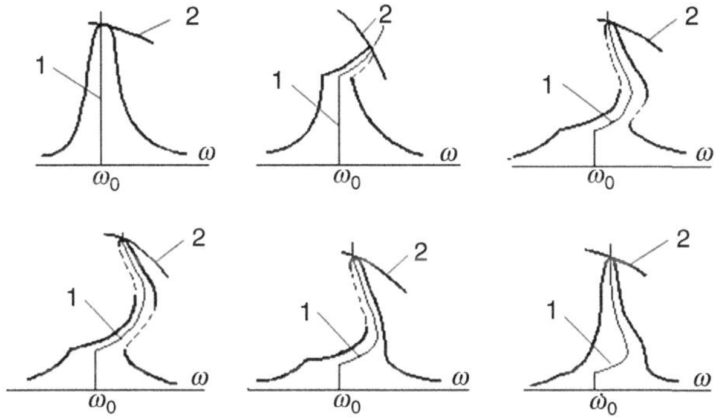

Fig. 7 The amplitude-frequency characteristics of system with cinematic drive for different values of driving speed

tuning are caused by the nonlinearity of the technological load. These difficulties can be eliminated by a transformation to a self-sustained vibration which is realized with the introduction of a circuit of positive feedback. With a certain tuning, called as autoresonance, at any variations of the parameters of the vibratory system and technological load, the mode of self_sustained vibrations with the maximum possible amplitude is conditioned automatically. The principal schematic of the autoresonant device for ultrasonic turning is shown in Fig. 1b. The power supply of the piezoelectric elements is carried out via a feedback path containing a feedback sensor 8, which records the vibration of one of the elements of the vibrating system; phase inverter Φ; and nonlinear amplifier 9. With sufficiently large gain on the initial part of the amplifier characteristic, the self-excitation of vibrations occurs, and the level of the limitation of the output voltage determines the amplitude of the steady state vibrations. If a phase shift in the feedback path is selected so that the phase difference between the vibrations of the cutter and the exciting force produced with a transducer corresponds to the resonance one, then the device realizes resonance vibration with the variation of the technology load and parameters of the vibrating system in wide ranges. Exactly this type of a system of exciting the vibrations with the resonance tuning of the phase in the feedback path is called autoresonance.

An important feature of such a system of vibration excitation should be noted. It consists of the fact that through a variation of the phase of the feedback, we can realize all of the amplitude–frequency characteristic of the nonlinear vibrating system, including unstable, and consequently unrealizable, branches during forced vibrations. The mentioned feature makes the autoresonant system steady state with respect to the significant variations of the parameters of the machine and nonlinear technological load. In traditional ultrasonic technological systems, the power supply is carried from a generator G with the independent excitation shown in Fig. 1 by dashed lines. In this case, even with the presence of tuning internal feedbacks in the generator, the ultrasonic system is capable of working only with low technological loads that cause no appearance of unstable branches of resonance curves.

Acknowledgements This work was supported by the Russian Foundation for Basic Research, project 09-08-00941-a.

References

1. Astashev, V.: The tuning of ultrasonic machines whilst under load. Stanki i Instrumenti (Machine-tools and Tools) 10. (in Russian) (1972)
2. Astashev, V.: The influence of ultrasonic vibration on plastic deformation processes. Machinovedenie (Machine Science) 2. (in Russian) (1983)
3. Astashev, V.: Effect of ultrasonic vibrations of a single point tool on the process of cutting. J. Mach. Manuf. Reliab. **5**(3), 81–86 (1992)
4. Astashev, V., Babitsky, V.: The efficiency of resonantly tuned ultrasonic machines. Machinovedenie (Machine Science) 6. (in Russian) (1981)
5. Astashev, V., Babitsky, V.: Methods of ultrasonic machine efficiency increase. Stanki i Instrumenti (Machine-tools and Tools) 3. (in Russian) (1982)
6. Astashev, V., Babitsky, V.: Ultrasonic cutting as nonlinear (vibro-impact) process. Ultrasonics **36**(1–5) (1998)
7. Astashev, V., Babitsky, V.: Ultrasonic Processes and Machines. Springer, Berlin (2007)
8. Astashev, V., Sakaian, A.: Experimental investigation of ultrasonic machine dynamics. Machinovedenie (Machine Science) 4. (in Russian) (1967)
9. Isaev, A., Anokhin, V.: Application of ultrasonically vibrating tool to metal cutting. Vestnik Machinostroenia (News in Machine Building) 5. (in Russian) (1961)
10. Izumi, O., Oyama, K., Suzuki, Y.: Effects of superimposed ultrasonic vibration during compressive deformation of metals. Trans. Jpn. Inst. Metals **7**(3) (1966)
11. Markov, A.: Ultrasonic Machining of Intractable Materials. Illife Books, London (1966)
12. Robinson, A.: The application of ultrasonic energy to metal wire drawing. Wire Wire Prod. **39**(12) (1964)
13. Rosenberg, L., Kazantsev, V., Makarov, L., Yahimovich, D.: Ultrasonic Cutting. Nauka, Moscow (in Russian) (1964)
14. Severdenko, V., Klubovich, V., Stepanenko, A.: Ultrasonics and Plasticity. Nauka and technika, Minsk (in Russian) (1976)

Experimental Studies of Thermoviscohyperelastic Behaviour of Filled Elastomers

T.A. Beliakova, E.V. Lomakin, and Yu.P. Zezin

Abstract The viscohyperelastic properties of the hydrogenated nitril-butadien rubber containing the nano-size particles of technical carbon are studied. The stress–strain curves and the relaxation diagrams of the material under conditions both of tension and compression are presented. For the description of the viscohyperelastic behavior of filled elastomers, the constitutive equations are proposed. These equations represent the experimental values of stresses as the sum of two parts: hyperelastic part and viscoelastic one. The hyperelastic part is defined by the hyperelastic potential. The viscoelastic one is described by the nonlinear analogue of the equations of linear theory of viscoelasticity. The method for the determination of the parameters and the material functions is proposed. The comparisons of experimental data with the results of theoretical predictions are presented.

1 Introduction

Rubbers, both natural and synthetic, are rarely used as raw materials. Instead they usually constitute the base of polymer composites which preserve the elasticity of rubber but demonstrate improved deformation and strength properties. This strengthening of polymer materials is gained by an infusion into the rubber compound of dispersed inorganic particles – the fillers. Better strengthening is achieved for filler particles of a smaller size. A similar effect is discussed in [1]. In particular, infusion of coarse particles into the rubber compound is shown to

T.A. Beliakova (✉) · E.V. Lomakin
Faculty of Mechanics and Mathematics, Moscow State Lomonosov University, Moscow, Russia
e-mail: tanbel@mech.math.msu.su; lomakin@mech.math.msu.su

Y.P. Zezin
Institute of Mechanics, Moscow State Lomonosov University, Moscow, Russia
e-mail: yuzezin@gmail.com

H. Irschik et al. (eds.), *Advanced Dynamics and Model-Based Control of Structures and Machines*, DOI 10.1007/978-3-7091-0797-3_4,

increase several times the Young modulus and the strength of the rubber. Use of nanoparticle fillers (such as carbon black) increases Young modulus and the strength limit up to ten times. In the meantime, the filled elastomers preserve the strain limit of the initial pure rubber. Therefore the use of nanoparticle insertions for rubber compounds allows one either to improve the deformation properties of the material, for the same filler admixture, or to achieve the required stiffness and the strength properties with a reduced fillers admixture.

In this work we investigate the hyperviscoelastic behaviour of the polymer composite with elastomeric matrix based on hydrogenated nitrile-butadiene rubber reinforced with carbon black N 110, with the mass content 50% [2]. The filler under consideration is characterised by a low dispersion of particle sizes (within 20–25 nm), with the specific surface $1.25 \cdot 10^5$ m^2/kg and with the genuine density 1,850 kg/m^3. The elastomer specimens were tested under the conditions of uniaxial tension and uniaxial compression. In addition stress relaxation tests were performed both under the tension and the compression. All experiments were carried out at room temperature. The proper constitutive relations are suggested describing hyperviscoelastic properties of the filled elastomers, and the results of the approximations are compared with the experimental data obtained.

2 Experimental results

The elastomeric material under consideration is investigated in a series of experiments under the condition of uniaxial tension. It is determined that the tested material demonstrates a significant strength under tensile loads and possesses a high strain limit of more than 500%. So high strain limits are specific for unfilled rubbers.

The experiments were carried out with the use of the testing equipment specially deviced for polymeric materials. The corresponding electromechanical device was designed in the Institute of Mechanics of Moscow State Lomonosov University and it allows to test low-strength materials under the conditions of constant strain rate, stress relaxation, and load-unload process. The strain rate can be varied in the range from 10^{-5} to 10^{-2} s^{-1}. The strain gauge system was used to measure the tensile load.

The uniaxial tension tests were carried out using the specimens with the gage length 40 mm. The tested specimens were cut from a 2 mm-thick plate of the material. The solid cylinder specimens of 10 mm in diameter and 12 mm in height were used for the tests under the compression.

The experimental diagrams for the material under consideration are presented in Fig. 1, for the case of the uniaxial tension, and in Fig. 2, for the case of the uniaxial compression. The stress–strain dependencies correspond to the temperatures 19°C, 50°C, 100°C, and 150°C.

Viscoelastic properties of the elastomer are investigated on the base of stress relaxation tests under both tension and compression conditions. In these experiments, the levels of strain were up to 0.6. Our experimental dependencies of the engineering

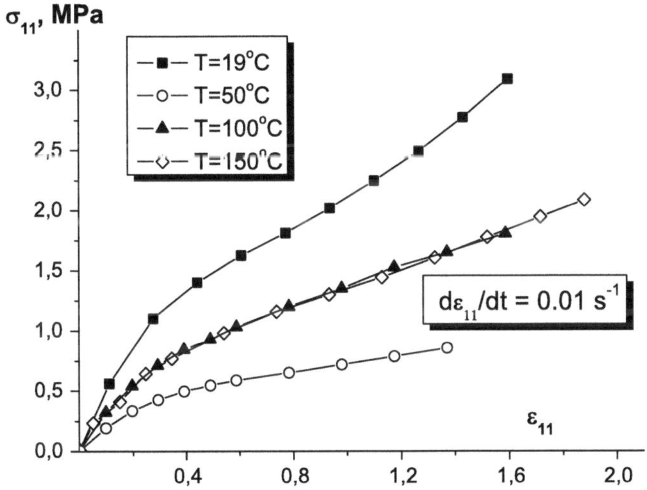

Fig. 1 Experimental stress–strain diagrams under the conditions of uniaxial tension

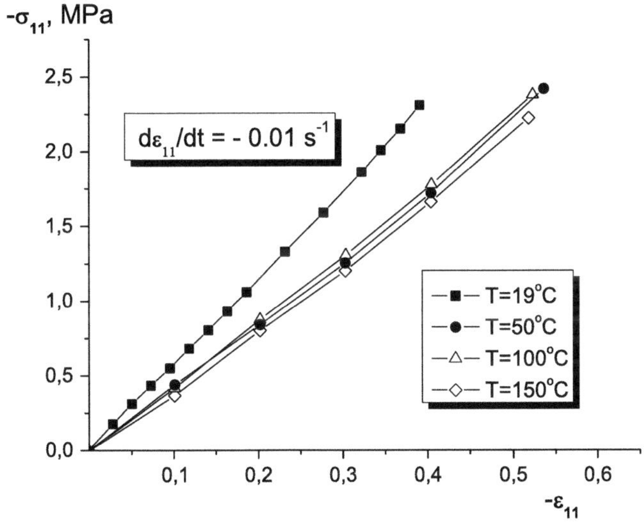

Fig. 2 Experimental stress–strain diagrams under the conditions of uniaxial compression

stress $\sigma_{11} = F/S_0$ (where F is the axial load and S_0 is the initial cross section area of the specimen) on time in the relaxation tests for the cases of tension and compression are shown in Figs. 3 and 4, respectively. In the relaxation tests under the condition of tension, the strain increases at the constant rate 0.01 s^{-1} up to the strain values ε_{11} =0.105; 0.185; 0.269; 0.507; 0.56 (E_{11} =0.111; 0.202; 0.305; 0.636; 0.717) where $\varepsilon_{11} = \lambda_1 - 1$ is the axial strain (engineering deformation), $E_{11} = (\lambda_1^2 - 1)/2$ is the Green–Lagrange axial strain, $\lambda_1 = L/L_0$, L_0 and L

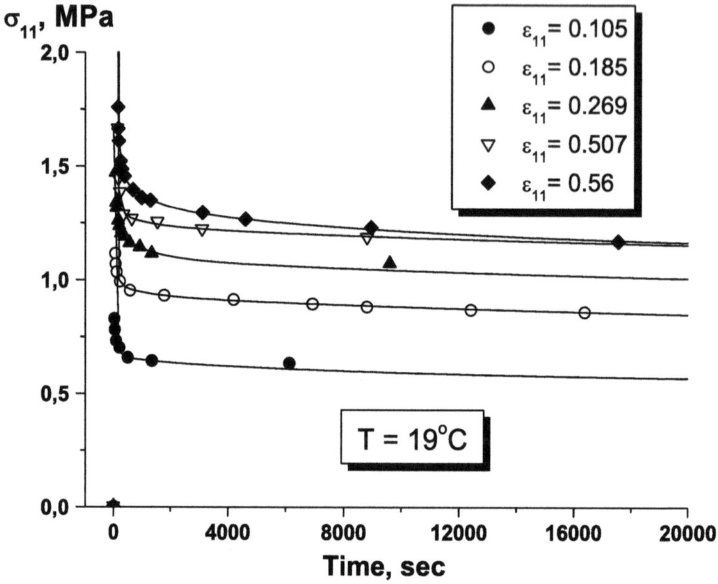

Fig. 3 Experimental engineering axial stress dependencies on the time at the different values of strain in tension

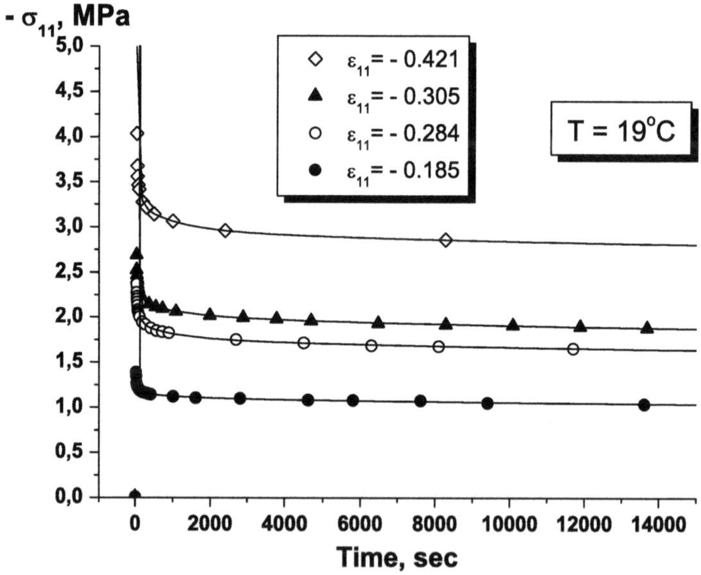

Fig. 4 Experimental engineering axial stress dependencies on the time at the different values of strain in compression

are the initial and the current lengths of the specimen gage part, respectively. The experiments for the stress relaxation in compression were carried out for the following values of the strain: $\varepsilon_{11} = -0.185; -0.284; -0.305; -0.421$ ($E_{11} = -0.168; -0.244; -0.259; -0.332$), $\lambda_1 = H/H_0$, where H_0 and H are the initial and the current heights of the specimen, respectively.

It can be seen from Figs. 3 and 4 that the results of the experiments reveal a physical nonlinearity of viscoelastic properties of the tested material. The values of relaxation modulus $\sigma_{11}/\varepsilon_{11}$ determined on the base of these diagrams depend on the strain level. An increase of the strain results in a decrease of the relaxation modulus, though this effect is more significant for the case of tension. Under the conditions of compression the values of relaxation modulus depends weakly on the strain level.

3 Constitutive relations

For the analysis of experimental data a generalised form of the constitutive relations for the hyperviscoelastic behaviour of the elastomer under consideration is used, which combines both equations of the nonlinear elasticity theory as well as the linear viscoelastic Boltzman-Volterra model. Thus, the components of the stress tensor can be presented as a sum of the hyperelastic and viscoelastic parts [3]:

$$\sigma_{ij} = \sigma_{ij}^h + \sigma_{ij}^v. \tag{1}$$

The hyperelastic components of the stress tensor σ_{ij}^h for an isotropic incompressible material can be obtained by differentiation of the elastic potential $W(I_1, I_2)$ with respect to an appropriate strain component. We take the elastic potential W in a polynomial form depending on the invariants $(I_1 - 3)$ and $(I_2 - 3)$ of the finite strain tensor [4]. In particular, the five-constant polynomial hyperelastic potential $W(I_1, I_2)$ can be written as:

$$W = c_{10}(I_1-3)+c_{01}(I_2-3)+c_{20}(I_1-3)^2+c_{11}(I_1-3)(I_2-3)+c_{02}(I_2-3)^2, \tag{2}$$

where the invariants I_1 and I_2 can be expressed through the axial elongations $\lambda_i = ds_i/ds_i^\circ$,

$$I_1 = \lambda_1^2 + \lambda_2^2 + \lambda_3^2, \quad I_2 = \lambda_1^2\lambda_2^2 + \lambda_1^2\lambda_3^2 + \lambda_2^2\lambda_3^2, \quad I_3 = \lambda_1\lambda_2\lambda_3.$$

For incompressible materials $I_3 = 1$.

The viscoelastic part of the stress tensor, in a general nonlinear form of constitutive relations, is taken as:

$$\boldsymbol{\Sigma}^v(t) = -p^v \boldsymbol{I} + \boldsymbol{F}(t) \int_0^t G(t-\tau) d\boldsymbol{E}(\tau) \boldsymbol{F}^T(t), \tag{3}$$

where $\boldsymbol{\Sigma}^v(t)$ is the viscoelastic Cauchy stress tensor, $F_{ij} = \partial x_i / \partial x_j^\circ$ are the components of the gradient tensor $\boldsymbol{F}$, with x_i° and x_i, $i = 1, 2, 3$, being the initial and the current coordinates, while $\boldsymbol{E}$ and $G(t)$ are the Green–Lagrange strain tensor and the relaxation function, respectively. In the case of small strains, the Green–Lagrange strain tensor coincides with the engineering one ($E_{11} \cong \varepsilon_{11}$) so that relations (3) turn into the constitutive relations of the linear theory of viscoelasticity:

$$\sigma_{ij}^{v}(t) = \int_0^t R(t-\tau)d\varepsilon_{11}(\tau). \tag{4}$$

The relation (4) is used to describe the viscoelastic components of stresses [5]. As one can see from the stress relaxation curves at Figs. 3 and 4, the hyperelastic part of stresses exceeds substantially the viscoelastic one. This justifies our use of the linear viscoelastic relations and guarantees that the error made is not large within the considered range of strains. The function of relaxation in relations (4) can be approximated by a sum of exponents:

$$R(t) = \sum_{i=1}^{N} E_i e^{-\alpha_i t}, \tag{5}$$

where the relaxation moduli E_i and the reduced times α_i are material-dependent constants.

Constitutive relations (1), (2), (4), and (5) imply that the quasistatic hyperelastic part of stresses and the viscoelastic part can be separated in description of a stress–strain state of elastomer materials. The viscoelastic part of stresses determines the dependence of the material deformation properties on the loading history. The stresses tend to their hyperelastic part under the conditions of slow loading. This is in agreement with the results of [3], where properties of elastomer materials under the conditions of deformation at high strain rates (about 10^3 s^{-1}) are discussed. In particular, the authors of paper [3] suggest that the hyperelastic part of the stress tensor should be determined from experiments at low (as compared to the values considered in this work) strain rates (0.001 s^{-1}) with the help of (1).

However, if the investigated strain rates are comparable with the experimental ones, another method can be used to determine the hyperelastic part of the stresses. The suggested approach is based on extracting asymptotic values of the axial stress from relaxation tests at different strain levels. Indeed, consider loading at a constant strain rate $\dot{\varepsilon}_{11}$, taking place for the time t_0, up to the strain level $\varepsilon_{11}^0 = \dot{\varepsilon}_{11} t_0$ and the subsequent stress relaxation process. On the base of (1), (2), (4), and (5) for this case, one can write for the axial stress in the specimen:

$$\sigma_{11}(t) = \begin{cases} \sigma_{11}^h(\varepsilon_{11}(t)) + \dot{\varepsilon}_{11} \sum_{i=1}^{N} E_i \dfrac{1-e^{-\alpha_i t}}{\alpha_i}, & 0 \le t \le t_0 \\ \sigma_{11}^h(\varepsilon_{11}(t)) + \dot{\varepsilon}_{11} \sum_{i=1}^{N} E_i \dfrac{e^{\alpha_i t_0}-1}{\alpha_i} e^{-\alpha_i t}, & t \ge t_0 \end{cases} \tag{6}$$

Table 1 Viscoelastic relaxation moduli for the approximation of experimental relaxation curves

ε_{11}^0	E_1, MPa	E_2, MPa	E_3, MPa	E_4, MPa	E_5, MPa
−0.421	0.422	0.444	0.664	0.918	4.568
−0.305	0.734	0.511	0.517	0.888	3.306
−0.284	0.583	0.485	0.565	0.789	2.427
−0.185	0.819	0.379	0.262	0.576	1.650
0.105	0.462	0.789	0.187	1.337	1.230
0.185	1.444	0.229	0.267	0.764	0.660
0.269	0.890	0.214	0.442	0.685	3.754
0.507	0.487	0.095	0.112	0.652	5.805
0.56	0.216	0.299	0.253	0.746	2.651

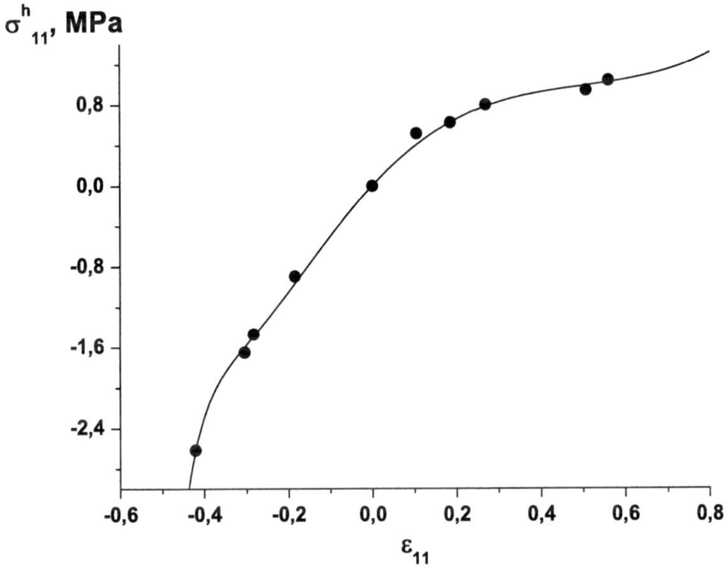

Fig. 5 Hyperelastic part of stress–strain diagram

The experimental data on stress relaxation presented in Figs. 3 and 4 can be approximated using the expressions from (6) with $N = 5$. The reduced times α_i are fixed to be: $\alpha_1 = 1.0\mathrm{e}^{-5}\,\mathrm{s}^{-1}$, $\alpha_2 = 1.0\mathrm{e}^{-4}\,\mathrm{s}^{-1}$, $\alpha_3 = 1.0\mathrm{e}^{-3}\,\mathrm{s}^{-1}$, $\alpha_4 = 0.01\,\mathrm{s}^{-1}$, and $\alpha_5 = 0.1\,\mathrm{s}^{-1}$. Then the relaxation moduli E_i are found by fitting the experimental stress relaxation diagrams (see Table 1). The resulting approximations for room temperature are shown as solid lines in Figs. 3 and 4.

Finally, the hyperelastic stress σ_{11}^h can be determined with the help of the hyperelastic potential (2). In particular, for room temperature, the material constants are found to be $c_{10} = 0.545$, $c_{01} = 0.204$, $c_{20} = 1.017$, $c_{11} = -2.420$, and $c_{02} = 1.041$ (the appropriate diagram is shown in Fig. 5).

4 Conclusions

The experimental studies of non-linear elastic and relaxation properties of filled elastomers are carried out. The stress–strain diagrams at constant strain rate under the conditions of uniaxial tension and uniaxial compression are obtained for different values of temperature. The relaxation properties are studied on the base of experiments of the deformation of specimens under constant strain rate up to certain strain levels and subsequent relaxation. A possible approach to the formulation of non-linear theory of thermoviscoelasticity for the characterization of elastomers behavior under the finite deformation is considered. The method for the determination of interrelated hyperelastic and rheological characteristics of deformation of filled polymeric materials is proposed, based on the long-term values of stresses during the process of relaxation. It is shown that good correspondence between the theoretical diagrams and experimental data can be achieved if the reduced times are fixed but the relaxation moduli are found by fitting the experimental stress relaxation diagrams. The hyperelastic stress–strain diagram is defined and the material constants in hyperelastic potential are determined. A quite satisfactory correspondence between the experimental data and the results of theoretical predictions is demonstrated.

Acknowledgements Grateful acknowledgment for support of this work is made to Russian Foundation for Basic Research (grants 09-08-00502, 11-01-00168).

References

1. Kraus, G.: Reinforcement of elastomers by carbon black. Rubb. Chem. Technol. **5**(4), 33–37 (1978)
2. Ivanovsky, V.I.: Technical Carbon. Processes and Equipment: Manual. Omsk, "Technical Carbon" (in russian) (2004)
3. Yang, L.M., Shim, V.P.W., Lim, C.T.: A visco-hyperelastic approach to modelling the constitutive behaviour of rubber. Int. J. Impact Eng. **24**, 545–560 (2000)
4. Rivlin, R.S.: Some topics in finite elasticity. In: Goodier, J.N., Hoff, N.J. (eds.) Proceedings of the 1st First Symposium Naval Structural Mechanics, pp. 169–198. Pergamon Press, NY, USA (1960)
5. Beliakova, T.A., Zezin, Yu.P., Lomakin, Eu.V.: Thermoviscohyperelastic behaviour of elastomeric materials filled with nano-particles. Mech. Solids **45**(4), 63–81 (2010)

Example of Instability in Drive Mechanisms

A.K. Belyaev

Abstract Stability of rotation of a drillstring in curved oil wellbores is studied. The drillstring is considered as a Cosserat rod and the equations for a spatially curved drillstring and hole are derived and studied in detail for circular cross sections. The cases which allow for the solution in closed form are analysed. The rotation with a constant angular velocity is shown to be feasible only for a straight drillstring. In all other cases the rotation is unstable and for some relations between the parameters it is accompanied by a quasi-static transition to a new configuration. The instability is caused by the energy exchange between the torsional and bending modes. The increasing torque on the end of the drillstring improves the rotational stability at expenses of the increasing stresses in the string.

1 Introduction

Drillstring failures caused by unstable rotation of the drilling assembly are continually ranked as one of the most frequent and costly problems in the oil industry. Although many other methods of oilwell drilling have been tried historically, rotary drilling technique dominates the industry. The drillstring components, i.e. drillpipes, are typically 5" in outside diameter and 4.275" in inside diameter. Depthwise the assemblies can extend up to 8 km or more downhole, i.e. drillstring is an extremely flexible slender rod. Directional drilling toward the desired target, which is very popular nowadays, complicates the problem as oil wells are drilled with substantial curvatures and three-dimensional characteristics. In order to provide pressure for

A.K. Belyaev (✉)
Institute for Problems in Mechanical Engineering, Russian Academy of Sciences,
V.O. Bolshoy pr.61, 199178 St.Petersburg, Russia
e-mail: vice.ipme@gmail.com

H. Irschik et al. (eds.), *Advanced Dynamics and Model-Based Control of Structures and Machines*, DOI 10.1007/978-3-7091-0797-3_5,

forcing the drillbit into formation, especially in the case of horizontal drilling, the drillstring is axially compressed, i.e. it is buckled in its static configuration and rotates in a curved well. The objective of the paper is to analyse the stability of the drillstring rotation.

The paper is organised as follows. The first part is concerned with the modelling of the string by means of the rod theory. The second part deals with the most important case, in which no friction is assumed, and in addition to this, the string and the wellbore have circular cross sections. The cases of the helical and circular initial forms of the string and the wellbore are studied in detail. It is shown that the rotation is unstable and for some relations between the parameters it is accompanied by jump transition to a new configuration. The influence of the torque at the hole bottom on the rotational stability and the string strength is studied in the last part.

2 Drillstring as a Cosserat line. Geometry and governing equations

The drillstring under consideration is considered as a rod, i.e. a 1D line. All the equations of the differential geometry, e.g. [1], are applicable to this line. In particular, the motion of the natural trihedron of a spatial curve is prescribed by the position of the generic point of a spatial curve. This position is given in the system of fixed axes by the position vector $\mathbf{r}$ which is considered as a prescribed function of a curvilinear coordinate s counted along this curve from the origin. The unit vector $\mathbf{t}$ tangent to the curve is given by $\mathbf{t} = d\mathbf{r}/ds$. The infinitesimal vector $d\mathbf{t}$ is perpendicular to $\mathbf{t}$ and directed to the concave side of the curve. For this reason the unit vector in the tangent plane $\mathbf{n} = \rho\, d\mathbf{t}/ds$ is referred to as the unit vector of the principal normal to the curve. Here ρ denotes the curvature radius. The unit vector of binormal is constructed due to the rule $\mathbf{b} = \mathbf{t} \times \mathbf{n}$. Thus, an orthogonal trihedron $\mathbf{t}, \mathbf{n}, \mathbf{b}$ is determined at any point of the curve. The Frenet formulae of the differential geometry, e.g. [1], are as follows

$$d\mathbf{n}/ds = \mathbf{D} \times \mathbf{n}, \quad d\mathbf{t}/ds = \mathbf{D} \times \mathbf{t}, \quad d\mathbf{b}/ds = \mathbf{D} \times \mathbf{b}, \tag{1}$$

where $\mathbf{D} = w\mathbf{t} + k\mathbf{b}$ denotes the Darboux vector, with the twist being designated by w, and $\times$ stands for the vector product. In what follows we assume that the tangent in the natural and actual configuration of the curve are coincident. Then we can introduce the generalised Darboux vector $\boldsymbol{\Omega} = \mathbf{D} + \mathbf{t} d\theta/ds$ with $\theta\,(s)$ denoting the angle of rotation of the cross-section about the tangent $\mathbf{t}$.

Equilibrium equations for a spatially curved rod are taken in the form of Kirchhoff's equations

$$d\mathbf{Q}/ds + \mathbf{q} = 0, \quad d\mathbf{M}/ds + \mathbf{Q} \times \mathbf{t} + \mathbf{m} = 0, \tag{2}$$

see [2]. Here **M** and **Q** denote respectively the moment and the force in the cross-section whilst **q** and **m** denote respectively the external distributed force and moment acting on the rod. As it will be shown below there is no need for ascribing a particular constitutive law for the force in the rod. In the actual configuration, the moments in the rod in projections on the principal axes 1 and 2 of the cross-section and tangent **t** are taken in the form of Clebsch's constitutive equations

$$M_1 = EI\left(\kappa_1 - \kappa_{10}\right), M_2 = EI\left(\kappa_2 - \kappa_{20}\right), M_t = GI_p\left(\kappa_t - \kappa_{t0}\right). \tag{3}$$

Here the subscript 0 designates the initial configuration of the rod (natural state), M_1 and M_2 are the bending moments and M_t is the torque. The bending and torsional rigidities are denoted respectively by EI and GI_p. Rod's curvatures are expressed in terms of twist w, the curvature radius ρ and angle θ

$$\kappa_1 = \sin\theta/\rho, \quad \kappa_2 = \cos\theta/\rho, \quad \kappa_t = w + d\theta/ds. \tag{4}$$

Equilibrium equation (2) for the moments in projections on the principal axes 1 and 2 of the cross-section and the tangent **t** are as follows

$$\begin{aligned} dM_1/ds + \kappa_2 M_t - \kappa_3 M_2 - Q_2 + m_1 &= 0, \\ dM_2/ds + \kappa_3 M_1 - \kappa_1 M_t + Q_1 + m_2 &= 0, \\ dM_t/ds + \kappa_1 M_2 - \kappa_{21} M_1 + m_t &= 0. \end{aligned} \tag{5}$$

In what follows we adopt the following assumptions: (1) drillstring is a homogeneous linear elastic rod, (2) the system exposes no material damping and no friction between the string and oil well, (3) quasi-static rotation is considered, and (4) the string is driven at its end $s = 0$ with a constant angular velocity.

Let the reference configuration of the drillstring be given by w_s, ρ_s and θ_s. The actual configuration is assumed to take the form of the well prescribed by w_p, ρ_p and θ_p. The third equation in (5), which is the equation for torque, is seen to be independent of the forces in the rod. In addition to this, $m = 0$ since no friction is assumed between the drillstring and the oilwell wall during the drillstring motion. Substituting (2) and (3) into the sixth equation in (5) results in the single differential equation

$$\frac{d^2\theta}{ds^2} - \frac{1+\nu}{\rho_s\rho_p}\sin\theta = \frac{d}{ds}\left(w_s - w_p\right), \tag{6}$$

where ν denotes Poisson's ratio. In the case of circular cross-sections of the drillstring and the well, the new quantity $\theta = \theta_p - \theta_s$ implies the angle of rotation of drillstring's cross-section relative to the natural state.

3 Free rotation of drillstring in a planar wellbore of the arc form

Equation (6) describing the quasi-static rotation can be simplified and made more transparent. First, for all forms of the drillstring and the well, for which $w_s = const$ and $w_p = const$, the right hand side of (6) vanishes. This is the case of a circular arc and a helix. Also the straight line belongs to this particular class. To begin with, we consider the case of rotation of a drill string of a circular form of the radius R_s in a circular well of the radius R_p. In this case, studied in [3], (6) takes the following form

$$\frac{d^2\theta}{d\varphi^2} - a^2 \sin\theta = 0\,, \quad a = \sqrt{\frac{R_p}{R_s}(1+\nu)} \tag{7}$$

and the angle $\varphi = s/R_p$ is the angular variable corresponding to the curvilinear variable s. The case of a helical drillstring rotating in a plane well of the arc form (6) reduces to (7) in which

$$a = \sqrt{\frac{R_p R_h (1+\nu)}{R_h^2 + (h/2\pi)^2}}, \quad w_s = \frac{R_h}{R_h^2 + (h/2\pi)^2}, \quad \rho_s = \frac{R_h^2 + (h/2\pi)^2}{R_h}. \tag{8}$$

Here twist w_s and the curvature radius ρ_s of the helical drill string are expressed in terms of the radius R_h and the height h of the helix.

Since both cases are described by the same differential equation (7) our consideration is reduced to this equation regardless of the particular form of the drillstring.

The general form of the boundary conditions to (7) takes the form

$$\varphi = 0\,, \quad \theta = \Theta_0\,, \quad \varphi = \Psi\,, \quad GI_p\left(w_p - w_s + {R_p}^{-1} d\theta/d\varphi\right) = -M_{t1}, \tag{9}$$

where Θ_0 denotes the angle of rotation of the driven end of the drillstring ($\varphi = 0$) and M_{t1} stands for the external torque at the bottom end of the drillstring ($\varphi = \Psi$) caused for instance by external friction of the facility attached to the drillstring's bottom.

As we consider the planar well and free rotation, then $w_p = 0$ and $M_{t1} = 0$. In order to reduce the number of the parameters of the problem we limit our consideration to the case of the planar circular drillstring, i.e. $w_s = 0$. The boundary value problem, (7) and (9) has the following closed form solution

$$\Theta_0 = 2\arccos\left(\cos\frac{\Theta_1}{2}\sin\left\{\mathrm{am}\left[a\Psi + \mathrm{F}\left(\frac{\pi}{2}, \cos^2\frac{\Theta_1}{2}\right), \cos^2\frac{\Theta_1}{2}\right]\right\}\right), \tag{10}$$

where am $[\ldots,\ldots]$ and F $(\ldots,\ldots)$ denote the Jacobi amplitude function and the elliptic integral of the first kind, respectively, and Θ_1 denotes the angle of rotation of the free end of the drillstring.

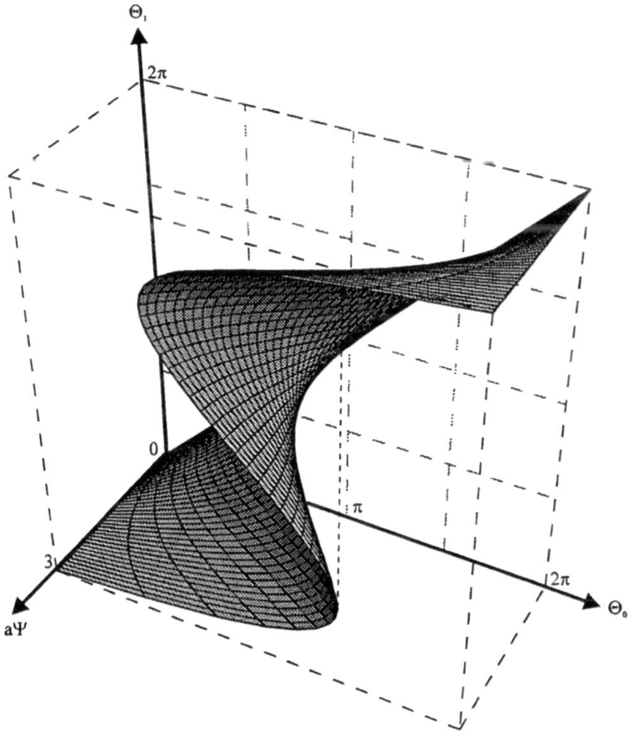

Fig. 1 Fold

Though the aim of the analysis is obtaining the dependence $\Theta_1 = \Theta_1(\Theta_0)$ the latter formula provides us with the inverse dependence. This form of the solution is preferable since it ensures a single-valued dependence. Figure 1 displays the dependence $\Theta_1 = \Theta_1(\Theta_0)$ versus the critical parameter a. A fold is observed which is an evidence of the nonlinearity of the problem and is known to be the cause of instability.

The sections of Fig. 1 are shown in Fig. 2 for some values of a. It is seen that the uniform rotation of the precurved drillstring in a curved well is not feasible since the case $\Theta_1 = \Theta_0$ occurs only for $a = 0$ as Fig. 1 shows. Then for $0 < a\Psi < \pi/2$ the angles of rotation of the drillstring ends are not coincident. Moreover, for $a\Psi > \pi/2$ there are several values of Θ_1 for certain values of Θ_0. This leads to the jumps which are marked by arrows in these Figures. For this reason, the solution in the form $\Theta_0 = \Theta_0(\Theta_1)$ is justified. The critical value of the parameter a is obtained from the equation $a_{crit}\Psi = \pi/2$, see Fig. 2, since this value separates the rotation with and without jumps.

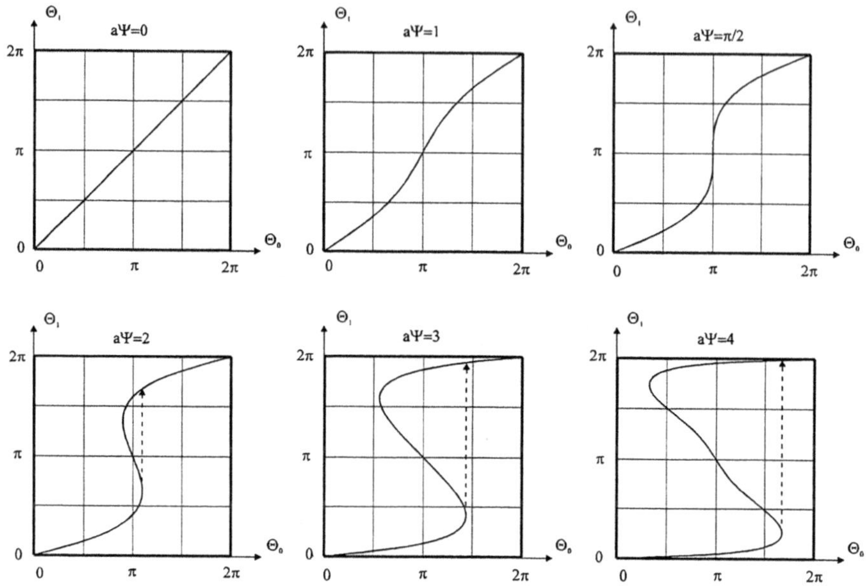

Fig. 2 Sections of the fold for some values of the critical parameter

4 The case of torque at drillstring's end

In this case $M_{t1} \neq 0$ and the closed form solution of the boundary-value problem, (7) and (9), is given by

$$\Theta_0 = 2\arccos\left(k\sin\left\{\mathrm{am}\left[a\Psi + \mathrm{F}\left(\varphi_1, k^2\right), k^2\right]\right\}\right), \tag{11}$$

$$\mathrm{k} = \sqrt{\xi^2 + \cos^2\frac{\Theta_1}{2}}, \quad \xi = \frac{w_s R_p}{2a}, \quad \varphi_1 = \arcsin\left(\frac{1}{k}\cos\frac{\Theta_1}{2}\right). \tag{12}$$

Figure 3 shows the angle of rotation Θ_1 vs. Θ_0 for $a\Psi = 2$ and some values of the non-dimensional torque $\lambda = R_p M_{t1}/2aGI_p$. One can observe that increase in the external torque improves the stability of rotation.

5 Torque at the rotary table

Equations (3) and (4) provide one with the formula for the torque in drillstring. The torque which is required for rotation of the drillstring, i.e. the torque of the rotary table, is as follows

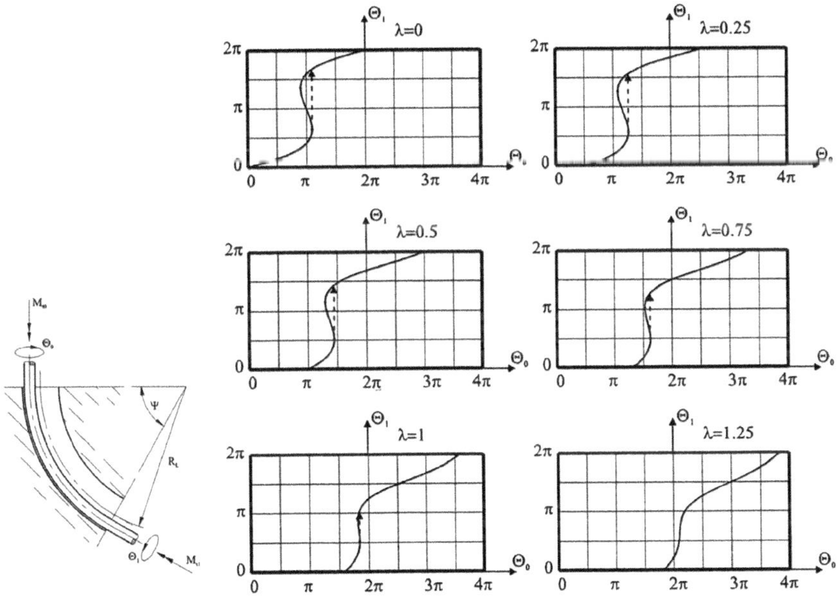

Fig. 3 The angle of rotation for some values of the non-dimensional torque

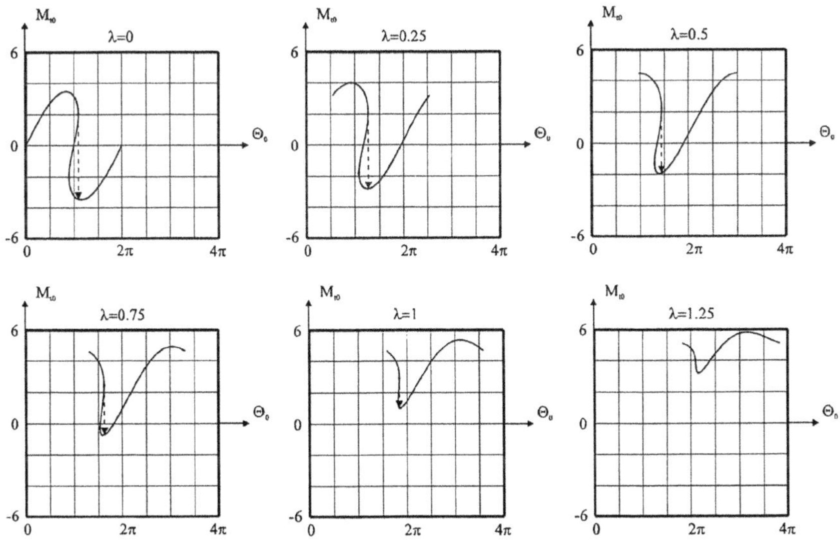

Fig. 4 The torque at the rotary table for some values of the torque at borehole's bottom

$$M_t = GI_p R_p{}^{-1} d\theta/d\varphi\big|_{\varphi=\Psi} = GI_p R_p{}^{-1}\sqrt{2\,(\cos\Theta_1 - \cos\Theta_0)}. \qquad (13)$$

Inserting (4) yields the dimensionless torque $M_t\ R_p/GI_p$ at the rotary table which is shown in Fig. 4 for several values of the torque λ at the bottom of the borehole.

6 Conclusions

The quasi-static stability of an arbitrary flexible elastic drillstring rotating in curved well is studied. To this aim, the drillstring is modelled by a Cosserat rod and the equations for a spatial curved drillstring are used. Their factors are shown to be expressed in terms of the curvatures of the deflected axes and the unit angle of twist in actual and reference configuration and the torsional and flexural rigidity of drillstring. In the most important case (the well and the drillstring have circular cross sections and no friction is assumed), these six equations are reducible to a singular equation for the rotation angle of the cross-section. Some cases allow for the solution in closed form in terms of the elliptic functions and integrals. The cases where the initial form of the drillstring is a helix and the well is a plane circular arc are analysed. It is shown that the rotation with a constant angular velocity is feasible only for a straight drillstring. In all other cases the rotation is unstable and for some relations between the parameters it is accompanied by a quasi-static transition to a new configuration. The increasing torque on the end of the drillstring improves the rotational stability at expenses of the increasing stresses in the drillpipes. The torque exhibits a behaviour which allows one to interpret the unstable rotation as stick-slip oscillations or chatter. However, neither external friction nor material damping is present in the model. The instability is caused by the energy exchange between the torsional and bending modes. The critical value of the torque is determined. The rotation becomes unstable when this critical value is exceeded. Local jumps of a buckled drillstring rotating in a curved oil wellbore is a well-known phenomenon which can be explained by means of the suggested approach, see e.g. [4]. The bending moments and the force factors in the drillstring are determined by means of the method of the successive approximations from (2). They are of crucial importance for directional drilling. For example, the drillstring jumps out of the oilwell if no axial force is applied. A number of problems remain to be tackled. First, the effect of the friction should be studied. Second, the problem of dynamic analysis of the jumps is of interest. Clearly, this challenging problem is associated with propagation of the bending-torsional waves in the drillstring and is much more difficult than the present quasi-static analysis.

References

1. Lurie, A.I.: Analytical Mechanics. Springer-Verlag, Berlin-Heidelberg (2002)
2. Eliseev, V.V.: Mechanics of Elastic Bodies (in Russian). Publishers of the St.P STU, St. Petersburg (1999)
3. Panovko, Ya.G., Gubanova, I.I.: Stability and Vibrations of Elastic Systems (in Russian). Nauka, Moscow (1967)
4. Belyaev, A.K.: Dynamics of a buckled drillstring rotating in a curved oil wellbore. In: Moon, F.C. (ed.) New Applications of Nonlinear and Chaotic Dynamics in Mechanics, pp. 169–171. Kluwer, Dordrecht (1999)

Positioning Systems: Global Versus Local

Fabio Casciati and Li Jun Wu

Abstract Current technology developments are addressed toward the wide class of bio-inspired achievements. Herds, flocks, shoals, all of them solve their relative positioning problems by vision schemes, which are presently requiring high technology and large storage masses. At a more affordable level, global positioning systems (GPS) are well established but they cannot be used indoor or within an urban environment. The number of visible satellites fluctuates from time to time and this is affecting the accuracy of any GPS receiver. Local positioning systems (LPS) are a valuable alternative. LPS is made of positioning sensors and at least four transponders which periodically transmit reference signals. The realization of a non-proprietary positioning sensor is sketched and discussed with focus on the levels of accuracy which can be achieved.

1 Introduction

Structural health monitoring (SHM), which diagnoses the health situation of structures according to their dynamic or static response, aims to improve the safety of human life and the performance of the societal arrangement. The data-base obtained from a SHM system can also be used to validate the design, study the dynamic characteristics of structures of new conception, and monitor the health situation of existing buildings.

Within SHM, a positioning system plays two different roles. First, the relative positioning, i.e., the displacement is a main physical variable from which the stress and the strain can be easily obtained. In this context, a global positioning system (GPS) sensor provides an absolute displacement measure at any time along any day

F. Casciati (✉) · L.J. Wu
Department of Structural Mechanics, University of Pavia, Pavia, Italy
e-mail: fabio@dipmec.unipv.it; lijun.wu@unipv.it

H. Irschik et al. (eds.), *Advanced Dynamics and Model-Based Control of Structures and Machines*, DOI 10.1007/978-3-7091-0797-3_6,

provided that an open vision of the sky is available. The second role results from the fact that many new kinds of sensor are introduced in order to monitor large-scale structures [1]. Mobile sensors are a viable solution in order to achieve a high density monitoring system by a limited number of sensors [2]. The position of each sensor however is a required information, easily achievable, for instance, by GPS.

High performance GPS sensors, i.e., geodesic GPS sensors, can measure the absolute displacement of the object point and achieve an accuracy of the order of centimeters [3, 4]. A GPS receiver is a non-contact sensor, which therefore can be easily installed. Signals from satellites cover a large range, but GPS sensors can only be effective in open sky. Furthermore, the geometry distribution of visible satellites fluctuates from time to time [5]. All these remarks affect the positioning accuracy from time to time and place to place. A valuable alternative would be the adoption of local positioning systems (LPS). One measures the distance from a transponder to the sensor by working on the carrier phase signal modulated by the so-called direct sequence spread spectrum (DSSS: indeed, signals from different satellites are modulated on the same radio frequency; in order to avoid conflicts, signals from different satellites are spread by different pseudo random noise (PRN) codes which are orthogonal each with the other). The carrier phase measurement is chosen due to its accuracy while the modulation is chosen due to its feasibility and extendibility [6].

After a quick review of the principles of high performance GPS (Principles of High Performance GPS), an open source receiver is discussed in "An Open Source GPS as Receiver Example". Two experiments are carried out. The GPS accuracy is introduced in "Collecting Data". In "The LPS Concept and its Challenges", the challenges encountered in designing the LPS system are analyzed and some approaches are introduced. Finally, the conclusions and a sketch of further future work are given.

2 Principles of high performance GPS

There are 32 satellites on orbit around the Earth (16 Feb 2011). Given the positions of the satellites, (x^i, y^i, z^i) for $j = 1, 2, \ldots, 32$ in an orthogonal coordinate reference system, the distance from each satellite to the ith sensor (in (x_i, y_i, z_i)) can be described by the relation:

$$r_i^j = \sqrt{(x^j - x_i)^2 + (y^j - y_i)^2 + (z^j - z_i)^2} \tag{1}$$

Theoretically, this distance can also be expressed as the product of the speed of the light and the signal propagation time Δt_i^j, i.e., $(T_i - T^j)$ where T_i is the time at which the GPS signal reaches the ith receiver, and T^j is the time at which the GPS signal leaves the jth satellite.

$$r_i^j = c\Delta t_i^j = c(T_i - T^j) \tag{2}$$

By measuring the signal propagation time from four satellites, one writes four equations which, once solved, provide the coordinates of the receiver (x_i, y_i, z_i) as well as the time difference between the satellite and receiver clocks. In order to acquire the most accurate signal arrival time, carrier phase measurement is adopted.

Since the receiver cannot be exactly synchronized with the satellite, the distance obtained for each sensor is approximate and called pseudo-range. Moreover, several factors, such as orbital errors, relativistic effects, ionosphere delay, troposphere delay, time inaccuracy, multipath, thermal noise and hardware bias, are affecting the real accuracy of the measured distance from the satellite to the sensor.

There are different technologies which can be employed to reduce the errors. For relativistic effects, ionosphere delay and troposphere delay, some correction models are available. In order to keep satellites synchronized each with the other, each satellite has its own cesium atomic clock. Simultaneously, there is a main control system (MCS) operating on the earth which determines and transmits the clock correction parameters to the satellites for rebroadcast in the navigation message. Differential GPS (DGPS) can further improve the accuracy by removing the part of the residual errors which are common to both the reference station and the GPS receiver, such as orbit errors, relativistic effects, ionosphere delay, troposphere delay and thermal noise [7]. The dominant error source in DGPS is the multipath error [8]. Multipath mitigation techniques range from antenna design to receiver architecture design and post processing of observables [9]. By now, the multipath mitigation remains an area of active research interest [9].

3 An open source GPS as receiver example

The market offers several GPS units, each with its own targeted applications. But to use proprietary items makes very difficult any progress achievable by basic research. A first step toward the construction of a suitable receiver is the availability of a hardware with open source software. This is the case of the GPS600 discussed in this section.

A block diagram of a GPS receiving system is shown in Fig. 1. The GPS receiver system consists of five principal components: antenna, receiver, processor, input/output (I/O) interface and power supply. Satellite signal, which is at radio frequencies (RF), is received via the antenna and then is down converted into intermediate frequency (IF) before its sampling as a digital signal. The digital signals are correlated in the correlator with an internally generated replica of the satellite code assumed. The results of the correlations are delivered to the processor which extracts the navigation message and controls the software signal tracking loops. The navigation message and tracking results are then processed by a positioning algorithm to obtain the position of the sensor: it is then sent to an input/output device for being displayed. Figure 2 is the block diagram of a typical open source GPS receiver based on GP4020. Figure 3 is the picture of the GPS600 based on GPS4020.

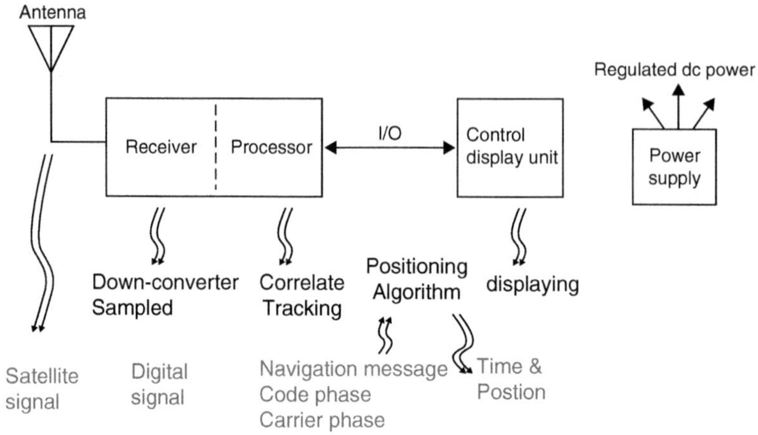

Fig. 1 Block diagram of a GPS receiver

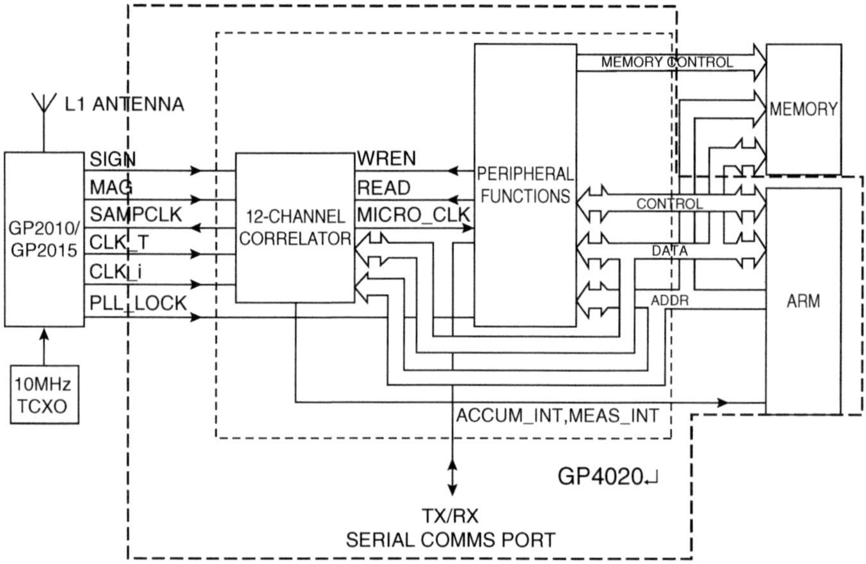

Fig. 2 Block diagram of the receiver

4 Collecting data

Currently, only the first of the several functions listed in Table 1 was uploaded in the GPS600. Then, the first experiment, in which a single static receiver is installed on the roof of a two floor building, is performed. The position of the receiver, which is provided in the Earth center Earth fixed (ECEF) coordinate (i.e., (x, y, z)), is

Fig. 3 Photograph of the GPS600

Table 1 Functions which could be uploaded on GPS600

Preliminary realized function	*Code phase measurement*
	Troposphere delay correction
	Ionosphere delay correction
	Carrier phase measurement
	Difference GPS

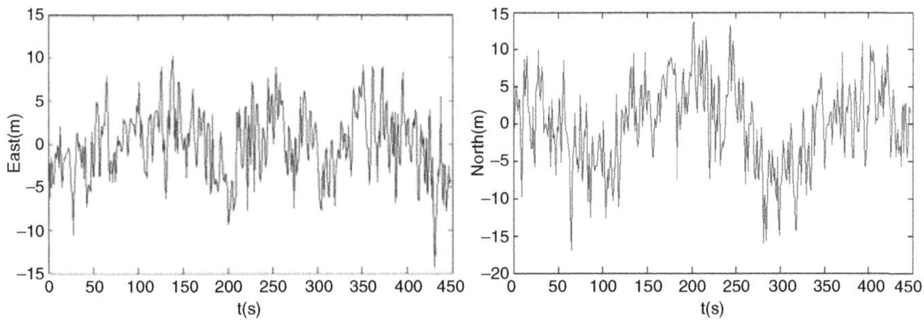

Fig. 4 Errors of East and North coordinates vs. time

projected onto the Earth's surface (i.e., (*North, East*)) by adopting the international 1924 model and the Gauss-Kruger projection algorithm. The errors on the North and East coordinates are shown in Fig. 4 while the nephogram is shown on Fig. 5. The 50% circular error probable (CEP) is 6.65meters. (CEP is defined as the radius of a circle which is centered on the mean: the boundary of this circle is expected

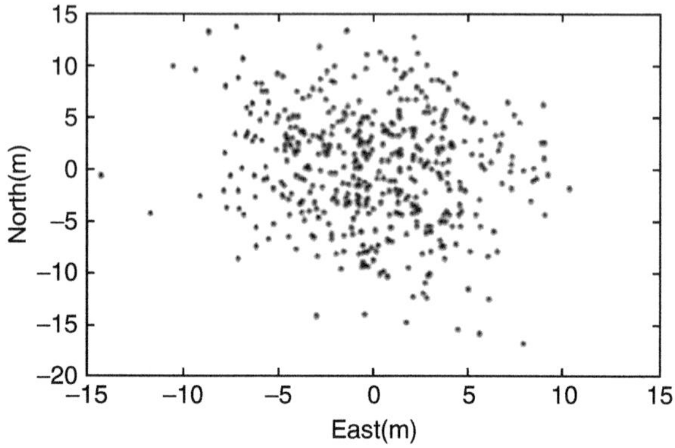

Fig. 5 North and East errors nephogram

```
←[HTime = 2010/6/6 13:20:0.365 (state:2)←[K
ECEF = (X:4.444623e+06 Y:7.146919e+05 Z:4.503368e+06) tb:1.678e-
02←[K
LLH  = (Lat:45.20292 Lon:9.13492 Hgt:143.17)←[K

State: positioning = 1, last position valid = 1, busy = 1
Ch: PN C PrV EpV Pseudorange    Elev. Azim.
 0: 18 L  1   1   2.875524e+07   53.8  61.0←[K
 1:  8 L  1   1   2.833927e+07  -17.6 338.8←[K
 2:  3 A  0   0←[K
 3:  7 C  0   0←[K
 4: 15 L  1   1   2.554950e+07    7.5  39.3←[K
 5: 28 L  1   1   2.699347e+07   -4.2 358.3←[K
 6:  4 A  0   0←[K
 7: 26 L  1   1   2.545897e+07   18.4  40.5←[K
 8: 10 C  0   0←[K
 9: 27 L  1   1   2.711031e+07   18.8  64.0←[K
10:  9 A  0   0←[K
11:  5 L  1   1   2.875029e+07  -38.4  76.1←[K
```

Htime	Collected time(according to the receiver clock)
State=2	The clock state.
ECEF	Earth center, earth fixed coordinate
x, y, z	The ECEF coordinate of the receiver (m)
LLH	Latitude (degree), longitude(degree), hight(m)
Positioning=1	The position is calculated (atleast 4 satellites are locked)
Last position valid=1	The calculation is valid (the solution converge to less than 0.1 meters)
Ch	Channel (each receiver has 12 channels to search and track satellite signals simultaneous-ly)
PN	Pseudo random number code of the received satellite in each Ch (channel).
C	Channelstate: A: acquisition C: confirm P: pull_in L: lock)
PrV	Pseudorange valid
EpV	Ephemeris valid (from the navigation data)
Pseudorange	The pseudorange
Elev.	Elevation (degree)
Azim.	Azimuth (degree)
K	Enter

Fig. 6 Single time data print from GPS600 recording (*left*). On the *right side* a reading help

to include 50% of the measurement.) This accuracy could be improved from three aspects before the carrier phase measurement is utilized:

1. Setting an elevation mask angle to mitigate multipath. The signal from a given satellite can reach the antenna by direct path or by reflected path (the so-called multipath, which greatly affects the accuracy [10]). Multipath arrives from angles near or below the horizon while direct path is characterized by larger incidence angle. As shown by the record in Fig. 6, collected by the GPS600, there are several signals from low elevations. Thence, setting an elevation mask angle is a necessary step toward an accuracy amelioration.
2. Adopting a Kalman filter. The current position algorithm utilized by the GPS600 is point solution, i.e., it does not carry any information from one measurement epoch to the next. Therefore, the solution accuracy is extremely dependent on the instantaneous satellites geometry which fluctuates from time to time. The Kalman filter can be adopted to address this problem [11]. In the Kalman filter

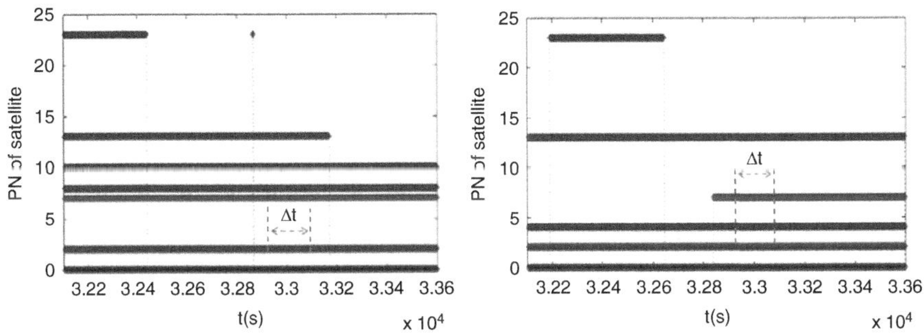

Fig. 7 Visible satellites distribution of receiver 1 (*left*) and receiver 2 (*right*)

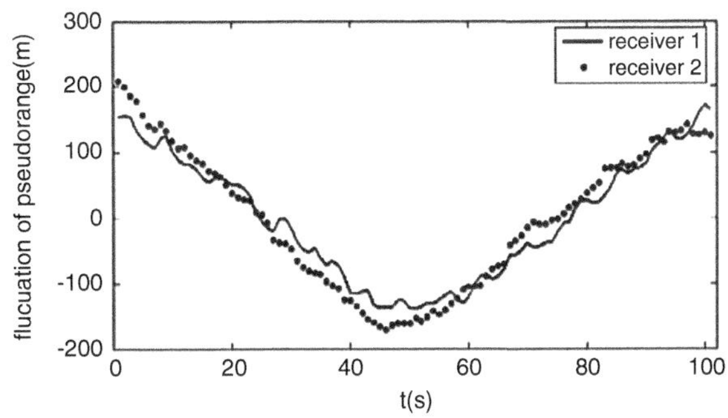

Fig. 8 Eight fluctuation of the pseudoranges from satellite #2

scheme, the updated state estimate is formed as a linear blend of the previous estimate(s) (projected forward to the current time) and the current measurement information.

3. Adopting the Difference GPS scheme. As previously referred, DGPS can improve the accuracy by removing the part of the residual errors which are common to both the reference station and the GPS receiver [7]. Instead of position coordinates, pseudoranges from each visible satellite to both receiver and reference station are differenced in most DGPS applications [12].

Therefore on the second experiment, data were collected by two receivers, installed on the roof of a two-floors building, in order to verify the correlation between the pseudoranges for the same visible satellite. Visible satellites distributions shown in Fig. 7 suggest selecting the two signals in Δt from satellite #2 as the analyzed signals. Instead of the linear trend of pseudoranges, attention is focussed on the fluctuations. Hence, the linear trends of the signals are first removed, after which the fluctuation of the two pseudoranges is drawn in Fig. 8. The fluctuations of the pseudoranges from receiver #1 and receiver #2 show high correlation. Hence, the accuracy of the rover receiver can be improved by computing the difference of the two pseudoranges.

5 The LPS concept and its challenges

By the available error reducing techniques, accuracy of geodesic GPS can achieve the order of centimeter. But the accuracy of GPS receiver relies on the number and the geometric distribution of the available satellites. Therefore, it cannot deliver equal precision in all position components at any time. This suggests to move to the study of Local Positioning Systems (LPS).

LPS utilizes several transponders at fixed positions as reference coordinates, just like GPS uses the satellites. The main idea of LPS is to deploy the transponders around the measuring sensor unit, which ensures that the measuring sensor can communicate with the transponders at any time. If the distances of the sensor from each transponder are determined, the sensor position then is obtained of consequence. Thus LPS can provide a stable positioning result.

There are several kinds of signals adopted for positioning system, such as ultrasonic, laser and electromagnetic waves. Ultrasonic signals will be greatly attenuated with the distance, which results in small cover range. Laser signals can only be effective in the line-of-sight, which means any obstacle can make them useless. For electromagnetic waves, there are several options, as ultra wide band (UWB), frequency modulated continue wave (FMCW), chirp spread spectrum (CSS) and measurement based on carrier phase signals modulated by DSSS [6]. Taking into consideration the cover range, the accuracy and the extendibility, the distance measurement technique based on signal propagation time, which is achieved by DSSS modulated carrier phase measurement, is preferred in the ongoing research effort [6]. Actually, there are several research groups adopting this carrier phase measurement in their positioning system [5, 13–15].

The transponders are mounted at known fixed positions on the surface of the earth in LPS. Thence, no ephemeris error, relativistic effects, ionosphere delay and troposphere delay are introduced in LPS [7]. The thermal noise jitter and the effects of interference induced measurement errors, introduced in the sensor, can be reduced by difference in a similar way as discussed for the GPS sensor [7].

The basic challenges include two aspects:

1. Multipath effect. Different from GPS, the dominating multipath for LPS is short multipath which is presently lacking effective mitigatory solutions. Recently, some promising methods were proposed to mitigate this short multipath, i.e., SMICLD (short-multipath insensitive code loop discriminator) [16].
2. Synchronization between different transponders. Since the electromagnetic signal propagates in the speed of light (290,000 Km/s), 1ms error in synchronization will induce 290 m error. Satellites in GPS system synchronize each with the other in high accuracy, and this is realized by the combination of cesium atomic clocks and main control system (MCS) on the earth. The atomic clock keeps the time accuracy while the control system, which is on the Earth, monitors and calculates the time difference between satellites. Within this framework, satellites can keep ns-order accuracy in their synchronization. But this method is too expensive for a LPS system. Hence, in current LPS systems, synchronization

are approached by wire, or by periodically broadcasting signal. More accurate methods would be welcome. By adopting the GPS signal generator GSS7790, the synchronization between different transponders could be solved. The GSS7790 offers the following advantages:

(a) It has multiple output signals which are synchronized each with the other, so the pseudolites could be deployed with the help of suitable transmission antennas.
(b) There is one combined GPS output available on the front panel for additional flexibility in view of regular receiver testing. Hence, it can be used to debug the receiver under development.
(c) The receivers can work under a configuration consistent with the outdoor configuration. Thus any receiver available on the market could be used for testing.

6 Conclusions and future work

Current technology developments are addressed toward the wide class of bio-inspired achievements. Herds, flocks, shoals, all of them solve their relative positioning problems by vision schemes, which are presently requiring high technology and large storage masses. This manuscript pursues the localization of a point in the space by GPS measurements. Three main issues are addressed:

1. Inside versus outside measurements
2. Proprietary receivers versus open source solutions
3. Global accuracy

Potential system architectures were discussed to move from the satellite visibility to measurements in roofed environment. A commercial module hardware, the GPS600 was acquired and some first measurements were carried out. A list of actions to be undertaken toward a better accuracy is also discussed.

Further future work includes two aspects: (1) Improving the performance of GPS600 by a serials of actions, for instance, setting elevation mask angle, realizing Kalman filter, utilizing DGPS, selecting the combination of visible satellites according to the geometry dilution of precision, realizing the ionosphere and troposphere corrections, realizing carrier phase measurement and improving multi-path mitigatory technique; (2) Deploying an LPS system based on the GPS signal generator and geodesic GPS receivers in order to test the idea of LPS.

Acknowledgements This research is supported by a grant from the Athenaeum Research Funds of the University of Pavia (FAR 2010). The research activity summarized in this paper was developed within the framework of the Marie Curie European project SMARTEN.

References

1. Chen, B.: A biologically inspired sensor network framework for autonomous structural health monitoring. Proc. SPIE **7292**, 729219 (2009)
2. Zhu, D., Qi, Q., Wang, Y., Lee, K.-M., Foong, S.: A prototype mobile wireless sensor net-work for structural health monitoring. In: Proceedings of SPIE, Nondestructive Characterization for Composite Materials, pp. 72941A-1 to 72941A-10. San Diego, CA, 8–12 March 2009
3. Casciati, F., Fuggini, C.: Monitoring an industrial steel building by GPS receivers. The 4th European Workshop on Structural Health Monitoring, pp. 219–226. DEStech, Lancaster, US (2008)
4. Casciati, F., Fuggini, C.: Engineering vibration monitoring by GPS: long duration re-cords. Earthquake Eng. Eng. Vib. **8**(3), 459–467 (2009)
5. Barnes, J., Van Cranenbroeck, J.: The Potential of a Ground Based Transceivers Network for Water Dam Deformation Monitoring. ICOLD, Barcelona, Spain (2006)
6. Casciati, F., Wu, L.J.: Investigating the potential of LPS in structural mechanics. The 10th International Conference on Motion and Vibration Control, Tokyo, Japan, 17–20 August 2010
7. Casciati, F., Wu, L.J.: Global vs local positioning systems. 5th World Conference on Structural Control and Monitoring, Tokyo, Japan, 11–14 July 2010
8. Counselma, C.C.: Multipath-rejecting GPS antennas. Proc. IEEE, **87**, 86–91 (1999)
9. Jardak, N., Samama, N.: Short multipath insensitive code loop discriminator. IEEE Trans. Aerospace Elect. Syst. **46**(1) (2010)
10. Casciati, F., Wu, L.J.: A high precision GPS receiver toward the introduction of LPS in structural mechanics. The 5th European Workshop on Structural Health Monitoring, Sorrento, Italy, 28th June–1th July 2010
11. Parkinson, B.W., Spilker, J., James, J.: Global Positioning System: Theory and Applications, vol. 1, pp. 424–429. American Institute of Aeronautics and Astronautics, Inc., Cambridge, Massachusetts (1996)
12. Kaplan, E.D., Hegarty, C.J.: Understanding GPS: Principles and Applications, 2nd ed. pp. 391–393. ARTECH HOUSE, INC., ISBN 1-58053-894-0 (2006)
13. Schwarzer, S., Vossiek, M., Pichler, M., Stelzer, A.: Precise distance measurement with IEEE 802.15.4 (ZigBee) device. IEEE Radio and Wireless Symposium, pp. 779–782. Orlando, FL, USA, 22–24 January 2008
14. Pichler, M., Schwarzer, S., Stelzer, A., Vossiek, M.: Positioning with Moving IEEE 802.15.4 (ZigBee) Transponders. IEEE MTT-S IMWS, Croatia (2009)
15. Pichler, M., Schwarzer, S., Stelzer, A., Vossiek, M.: Multi-channel distance measurement with IEEE 802.15.4 (ZigBee) devices. IEEE J. Selected Topics Signal Proc. **3**(5), (2009)
16. Vervisch-Picois, A., Samama, N.: Interference mitigation in a repeater and pseudolite in-door positioning system. IEEE J. Selected Topics Signal Proc. **3**(5), 810–820 (2009)

Real-Time Multi-channel Cable Replacement for Structural Control

Sara Casciati, Lucia Faravelli, and ZhiCong Chen

Abstract The goal of structural control is to mitigate the response of a structural system under the ongoing external excitation, on the basis of the feedback provided by a suitable array of sensors. In this study, a wireless sensing system, initially conceived for structural monitoring, is modified and used into a structural controller designed a reduced scale three storey steel frame: the goal is to replace the analog cables between the sensors and the structural controller. The wireless feedback to structural control comes from four accelerometers; the structural controller then drives an Active Mass Damper (AMD) actuator. The wireless sensing system is based on the recent low-cost System-on-Chip (SoC) radio transceivers instead of the commonly adopted commercial wireless modems. Unlike the structural monitoring system, a structural control system usually requires continuous and real-time sensor feedback which is implemented in this study by adopting the Frequency Division Multiplexing (FDM) method.

1 Introduction

Structural control systems aim to mitigate the vibrations of a structure, induced by external excitations, such as earthquake and wind load [6]. Unlike the structural monitoring applications, where periodic sensing with a low duty cycle (such as 2 min per day) is adequate and a high communication latency is acceptable, the structural control applications require a real-time and continuous sensing system, since the external excitation to the structure cannot be foreseen and the structural control system should react immediately [3]. This requirement represents

S. Casciati (✉)
Department ASTRA, University of Catania, via Maestranza 99, 96100 Syracuse, Italy
e-mail: saracasciati@msn.com

L. Faravelli · Z. Chen
Department of Structural Mechanics, University of Pavia, Via Ferrata 1, 27100 Pavia, Italy
e-mail: lucia@dipmec.unipv.it; zhicong.chen@unipv.it

H. Irschik et al. (eds.), *Advanced Dynamics and Model-Based Control of Structures and Machines*, DOI 10.1007/978-3-7091-0797-3_7,

a challenge for a sensing solution based on a wireless data transmission, which is broadly conceived to replace the expensive and labor-intensive cabled connections. The other challenges consist of the ability to cover the whole structure and the need for power supply sources of the wireless sensors [2, 4].

As a result of the recent development of the wireless communication and Integrated Chip (IC) technology, at present there is the commercial availability of a single chip wireless transceiver, which integrates most radio frequency components into a single chip and only requires a few peripheral passive components. In particular, the System on Chip (SoC) wireless transceiver also integrates a microcontroller core and many useful peripheral functional modules, such as memory, timer, analog to digital convertor, etc. The products of Texas Instrument (e.g., CC1020, CC1101, CC1110, CC2500, CC2510, CC2420, CC2430 etc.) provide typical examples of these devices and their fast evolution over time. The adoption of the most recent single chip wireless transceivers enables to easily implement, at a low cost, a customized wireless solution which can be targeted to any specific application. For this reason, the authors resorted to this technology and developed the herein described system architecture, which is tailored to structural control and health monitoring applications. It is worth mentioning that more expensive commercial wireless modems [7] are currently standardized for such applications and they offer a high computational capability, which, however, implies an increased effort by the end-user and limits the flexibility, especially in those applications where the computational tasks are not performed at the sensors level.

As mentioned in the first paragraph, the wireless sensing system for structural control applications should perform a real-time and continuous data transmission. The commonly used Time Division Multiplexing (TDM) approach would, instead, result into a significant network delay, especially when many sensors are deployed [5]. For this reason, a different approach is pursued by implementing the Frequency Division Multiplexing (FDM) method in the proposed wireless communication system. The FDM can be easily achieved by adjusting the programmable channel filter and the frequency synthesizer of the selected transceivers. The drawback of FDM consists of requiring a dedicated receiver for each single sensor node, since the transceiver cannot operate at several frequency channels at the same time. Nevertheless, the very low cost of the adopted single-chip transceivers, especially when they are purchased in large amounts, is such that the increased number of base station units does not imply a high increase of the total cost, which remains still more competitive than the one of standard modems.

In the following section, the wired structural control system which was implemented in the authors laboratory to reduce the vibration amplitudes of a 3-story steel frame during the motion of a small-size shaking table supporting the structure is briefly described. Subsequently, the modified wireless sensing system is introduced to replace the analog cables and the interface between the sensors and the structural controller is described. Finally, the results of a simple experiment for system validation are reported.

2 Structural control system

The typical active structural control system shown in Fig. 1 consists of the sensors which measure the response of the controlled structure, a controller which processes the data from the sensors according to the associated algorithm and determines the control output signal, and an actuator which receives the signal from the controller and then generates the control force to be applied to the structure. Therefore, if the controlled structure is regarded as a close loop system, the structural control system represents the feedback loop. In the authors' laboratory, a 3-storey steel frame of reduced size is placed on a shaking table whose motion is used as excitation. The response of the structure is measured by uniaxial accelerometers, Kinemetrics FBA-11, on each floor and at the base level. The actuator is an Active Mass Damper (AMD) installed at the top of the steel frame. The controller was designed and implemented as fulfillment of a doctoral study [1] and its details are summarized in the present section.

The structural controller features four analog inputs, one Serial Peripheral Interface (SPI) digital input and one analog output. In addition, a serial port compatible with the EIA-232 standard is available and it is used as interface when a command input is prompted from a PC or from any other compliant device. The core is a C8051F007 microcontroller produced by Silicon Laboratories. It is an 8051-based microprocessor equipped with many on-chip peripherals. The maximum clock frequency is 25 MHz, which is particularly suitable to process signals whose spectrum spans from 0 to 25 Hz. It has an in-system-programmable flash memory of 32 Kbytes and an additional on-chip RAM of 2 Kbytes. The microcontroller has a JTAG debug interface which supports on-line debugging and firmware downloads, which greatly facilitate its development.

The firmware has been designed with the easiness of use as the main concern. It has been implemented as a terminal shell. On the computer, a software, such as the one named HyperTerminal, can be used to access the embedded shell interface exposed by the board. The on-chip firmware for the wired structural control system is equivalent to the block diagram in Fig. 2.

The four analog inputs, which have a 12-bit resolution, are first passed through programmable amplifiers and programmable filters. The gains of the amplifiers can be programmed by setting a proper matrix. The filter has four options: (1) no filter; (2) offset cancellation; (3) approximation integration; (4) approximation integration followed by offset cancellation. The values obtained after filtering are the actual inputs to the controller.

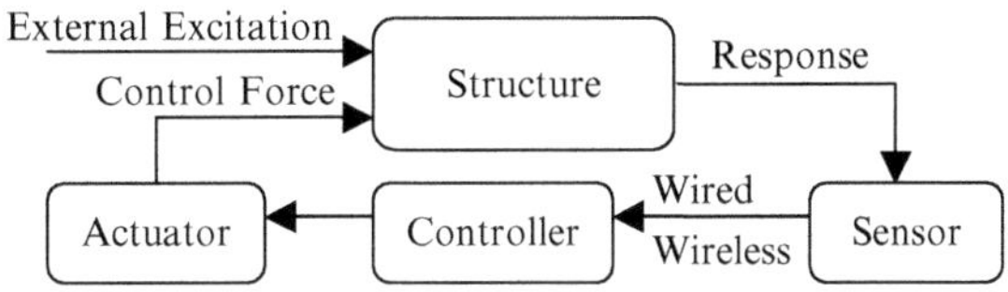

Fig. 1 The active structural control system

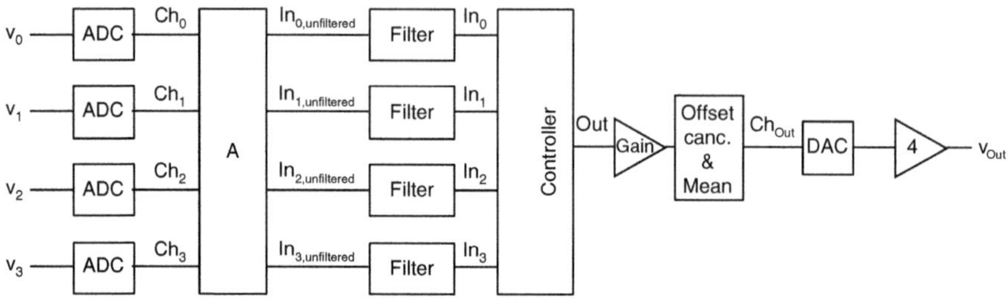

Fig. 2 The firmware block diagram for a wired system

The type of controller is user-selectable. The available controllers are: (1) no controller; (2) sinusoidal controller; (3) linear controller; and (4) fuzzy controller. The output of the controller is then amplified by a programmable factor and sent out through the 12-bit digital-to-analog converter. The desired mean value of the output channel is also user-selectable.

3 Wireless sensing system

For the wireless structural control system (Fig. 3), the ADC and analog signal input of the wired system are replaced by an SPI transferring the data to a wireless station unit. The user can select to either adopt the wired analog input or the wireless digital input using the shell interface.

3.1 System architecture

The wireless sensing system is conceived to replace the cable connection between the sensor and the structural controller. The Frequency Division Multiplexing (FDM) method is applied to ensure the real-time feature of the multi-channel data transmission. Indeed, when different channels operate on different frequency bands, the data transmission can occur simultaneously without conflicts. As shown in Fig. 3, each channel of the wireless sensing system is formed by a pair of transceivers, one of which is mounted on the wireless sensing unit and the other is connected to the structural controller by the SPI bus. In the SPI bus, the structural controller is the master end and the other transceivers are the slave ends. Only the structural controller can issue communication. In one cycle of remote data sampling, the controller broadcasts a command to the transceivers which then send the wireless request to the corresponding Wireless Sensing Units (WSUs). When the request is received, the WSUs start to perform AD conversion, and they send the sampled data back to the corresponding base station transceiver. In this manner, the synchronization of the sensing system is guaranteed. To achieve the sampled data, the structural controller communicates with the transceivers in turns, during predetermined time intervals.

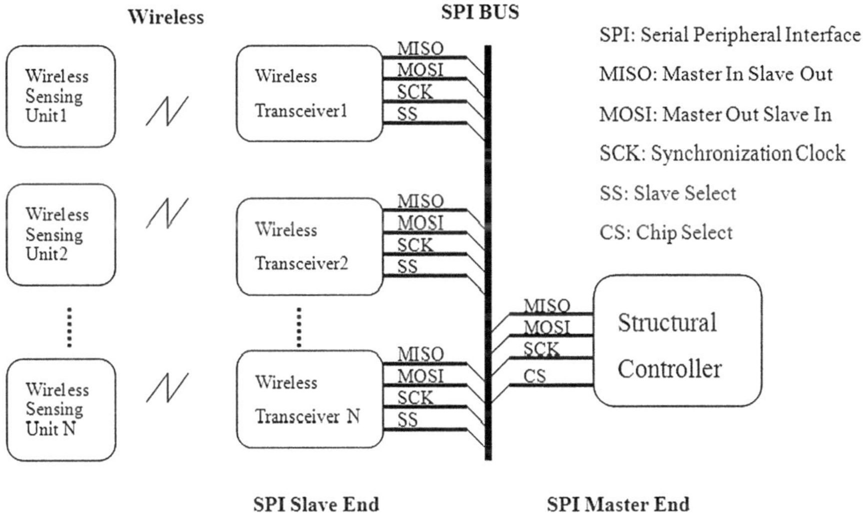

Fig. 3 Architecture of the wireless sensing system designed for structural control applications

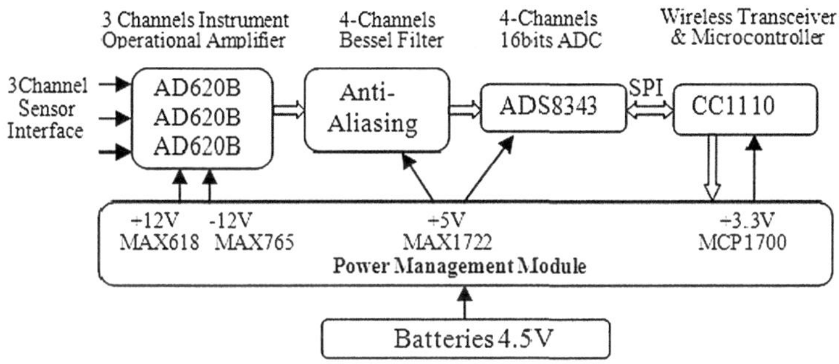

Fig. 4 The components-level block diagram of the wireless sensing unit

3.2 *Wireless sensing unit*

The wireless sensing unit associated to each sensor is installed on the structure and it is responsible for powering the sensor, acquiring the structural response data, and sending them to the wireless station. Therefore, it plays the most important role in the wireless sensing system. The block diagram of its hardware at the components-level is shown in Fig. 4.

Being the wireless sensing unit initially designed for a tri-axial accelerometer, three input channels are required. The low-power instrument AD620B is used to amplify the signal from a sensor and to change its bias in order to match the input to the analog-to-digital converter ADS8343. Before the conversion, a fourth order

Bessel filter of four channels is applied to perform the anti-aliasing. The ADS8343 has four channels, a resolution of 16 bit, and its maximum sampling rate is up to 100 kHz, which is sufficient for structural monitoring applications. The CC1110 is the previously described System on Chip (SoC) transceiver which includes not only a wireless transceiver, but also an enhanced 8051 microcontroller core, so that no external microcontroller is required.

In order to make this platform suitable for both low-power and non low-power structural monitoring applications involving different types of sensors, a flexible and efficient power management module is designed. It consists of a Low Dropout (LDO) linear regulator, MCP1700, with a quiescent current of 1.6 μA, and three switching regulators (namely, MAX618, MAX765, and MAX1722), which feature low quiescent current, highly efficient power conversion, adjustable output voltage, and a medium output power.

4 Experimental validation

In order to validate the performance of the proposed wireless sensing system, a simple laboratory experiment is carried out. The test setup is shown in Fig. 5. In this experiment, the acceleration records of a small-size, three-story steel frame under impact excitation are used as inputs to the controller board, which generates as output the displacement time history that should be assigned to the Active Mass Damper (AMD) located at the top of the structure. The signals simultaneously recorded by the three accelerometers located on each floor are transmitted to the

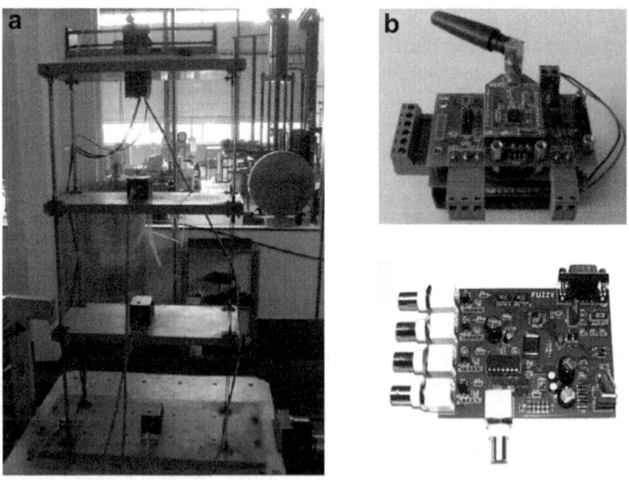

Fig. 5 Laboratory test setup: (**a**) monitored structure with AMD at *top (left)*; (**b**) wireless communication system between the sensors and the controller board

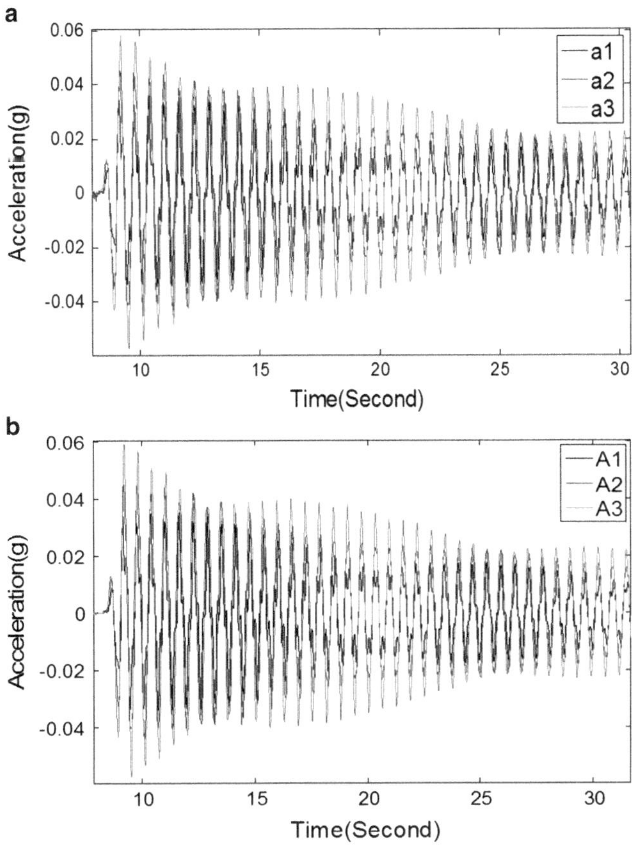

Fig. 6 Segments of the signals from the three accelerometers (inputs) acquired by (**a**) wireless connections; and (**b**) cable connections

controller board by using both the wireless and the cable connections. Segments of the acquired signals are separately plotted in Fig. 6. The sampling rate of analog to digital conversion is 50 Hz. In Fig. 7, enlarged segments of the signals in Fig. 6a, b are plotted together to compare the performances of the two data transmission systems. Namely, the signals from a1 to a3 correspond to cable connections, and the signals from A1 to A3 correspond to the wireless connections. In Fig. 8, segments of the output control signal computed with the input signals from the wired and wireless connections, respectively, are plotted together. From these figures, it is shown that a basically consistent agreement between each pair of plots is obtained.

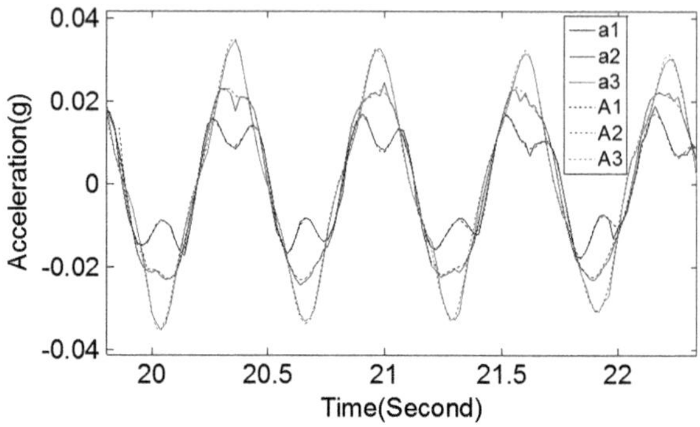

Fig. 7 Comparison of the input signals acquired by wireless and cable connections for an enlarged segment of the signals in Fig. 6

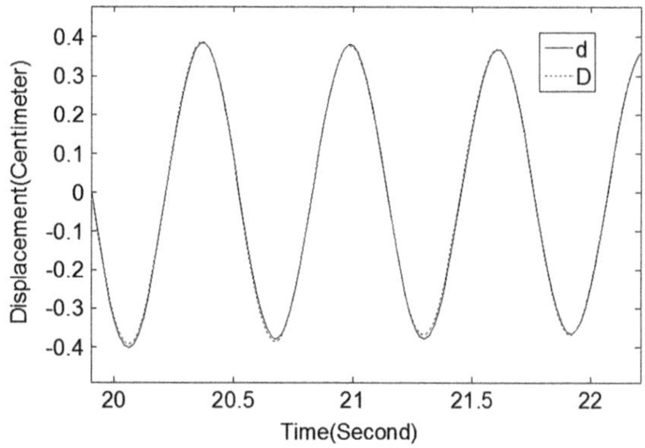

Fig. 8 Comparison of the controller outputs as computed from the wired (d) and wireless (D) input signals of Fig. 7

5 Conclusions

In the applications of structural control, continuous and real-time sensing techniques are required. Wireless connections between the sensors and the structural controller are desirable in order to reduce the high cost and the labor-intensive installation related to cables. In the present study, the analog cables are replaced by a customized prototype of a wireless sensing system which is based on the recent SoC single-chip transceivers. This solution offers high performance at low cost. The continuous and real-time wireless transmission is achieved by implementing a Frequency Division Multiplexing (FDM) approach. The overall system is validated by carrying out a laboratory experiment.

Acknowledgements This research is supported by the Athenaeum Research Funds from both the University of Catania (PRA 2007and PRA 2008) and the University of Pavia (FAR 2009). The research activity summarized in this paper was developed within the framework of the Marie Curie European project SMARTEN.

References

1. Battaini, M.: Controlled structural systems: design and reliability. Struct. Cont. Health Monit. **6**(1), 11–52 (1999)
2. Casciati, F., Faravelli, L., Rossi, R.: Architecture optimization for wireless sensor networks. In: Proceedings SPIE Smart Structures and Material Conference, vol. 5759. SPIE, Bellingham, WA, USA (2005)
3. Casciati, F., Rossi, R.: Fuzzy chip controllers and wireless links in smart structures. SMART03, Jadwisin, Poland, 2–5 September 2003
4. Casciati, F., Rossi, R.: A power harvester for wireless sensing applications. Struct. Contr. Health Monitor. **14**(4), 649–659 (2007)
5. Casciati, S., Faravelli, L., Chen, Z.C.: Design of a multi-channel real-time wireless connection system for analog cable replacement application. 7th International Workshop on Structural Health Monitoring, Stanford, CA USA, 9–11 September (2009)
6. Lynch, J.P., Wang, Y., Swartz, R.A., Lu, K.C., Loh, C.H.: Implementation of a closed-loop structural control system using wireless sensor networks. Struct. Contr. Health Monitoring **15**(4), 518–539 (2008)
7. Nagayama, T., Rice, J.A., Spencer, Jr. B.F.: Efficacy of Intel's Imote2 wireless sensor platform for structural health monitoring applications. Proceedings of the Asia-Pacific Workshop on Structural Health Monitoring. Yokohama, Japan (2006)

Flatness-Based MPC and Global Path Planning Towards Cognition-Supported Pick-and-Place Tasks of Tower Cranes

Markus Egretzberger, Knut Graichen, and Andreas Kugi

Abstract This paper proposes an overall control strategy towards cognition-supported pick-and-place tasks of tower cranes, as e.g. used in container handling applications. The presented concept is based on a global path planning algorithm which is combined with a flatness-based feedforward and a model predictive feedback control (MPC). A tower crane serves as an illustrative example for the application of the method. Assuming that the information about possible obstacles within the work space is provided by computer vision the concept is validated for pick-and-place tasks of the crane within the constrained work space.

1 Introduction

The importance of vision based concepts in closed-loop control has steadily increased over the last years, in particular in applications where an interaction with the environment plays a crucial role. Typical examples are autonomous vehicles and obstacle avoidance problems as well as (mobile) robots, see, e.g., [1, 2] for an overview. Most of the research works dealing with vision based control focus on the visual data acquisition and processing (cognition), while the actual control task is tackled with rather simple control strategies. Therefore, the main emphasis of this paper is placed on combining a global path planning strategy together with modern (nonlinear) open- and closed-loop control strategies. An interesting and

M. Egretzberger (✉) · A. Kugi
Automation and Control Institute, Vienna University of Technology, Austria
e-mail: egretzberger@acin.tuwien.ac.at; kugi@acin.tuwien.ac.at

K. Graichen
Institute of Measurement, Control and Microtechnology, Universität Ulm, Germany
e-mail: knut.graichen@uni-ulm.de

H. Irschik et al. (eds.), *Advanced Dynamics and Model-Based Control of Structures and Machines*, DOI 10.1007/978-3-7091-0797-3_8,

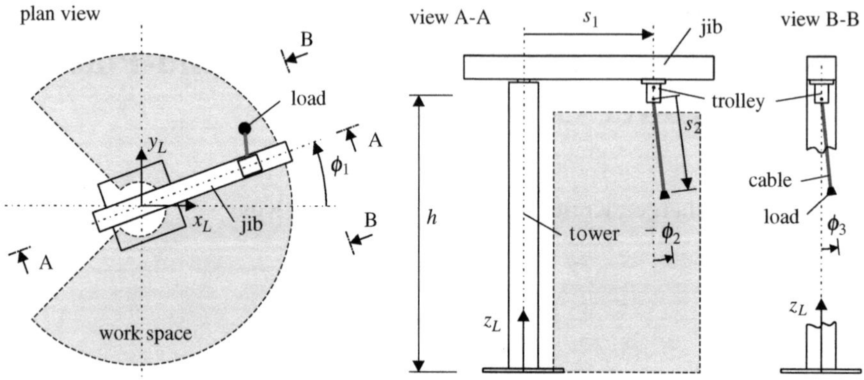

Fig. 1 Schematics of the tower crane

industrially important application for this type of combined concepts can be found in cognition-based pick-and-place tasks for tower cranes. The proposed overall path planning and control concept will be demonstrated by means of a tower crane considering point-to-point motions of the load while accounting for arbitrary obstacles within the work space.

A tower crane as schematically depicted in Fig. 1 is a typical example for a nonlinear, underactuated mechanical system. The crane as will be considered within the context of this paper consists of a fixed tower of the height h, a rotating jib with a movable trolley and a cable with an attached load. The motor driven jib rotates about the vertical axis of the tower with the angle ϕ_1. The radial motion of the trolley driven by a belt system is measured from the tower axis to the trolley's center of mass in terms of the translational degree-of-freedom (DOF) s_1. A cable winch fixed to the movable trolley together with a gimbal allow for the vertical actuation and an arbitrary pendulum motion of the cable with the attached load of mass m_L. The vertical motion is described in terms of the translational DOF s_2 measured from the gimbal to the load, while the rotation of the gimbal is considered as a composition of two successive rotations with the rotational DOFs ϕ_2 and ϕ_3. In summary, the crane possesses five DOFs with three actuated axes (jib rotation ϕ_1, trolley position s_1, cable length s_2) and two unactuated axes (gimbal angles ϕ_2, ϕ_3) allowing for an arbitrary positioning of the load within a hollow-cylindrical work space.

In the following Section 2, the mathematical model of the tower crane is derived before the combined concept of path planning and control is introduced in Section 3. The paper is concluded with some illustrative measurement results presented in Section 4.

2 Mathematical model

The equations of motion of the crane can be derived by means of Lagrange's formalism utilizing the generalized coordinates $\mathbf{q}^{\mathrm{T}} = [s_1, s_2, \phi_1, \phi_2, \phi_3]$ and generalized velocities $\dot{\mathbf{q}}^{\mathrm{T}} = [\dot{s}_1, \dot{s}_2, \dot{\phi}_1, \dot{\phi}_2, \dot{\phi}_3]$. Thereby, the three actuated axes are controlled by sufficiently fast velocity controllers such that the velocities $\dot{\mathbf{q}}_A = [\dot{s}_1, \dot{s}_2, \dot{\phi}_1]$ can be considered as the actual control input $\mathbf{v} = \dot{\mathbf{q}}_A$. On the assumption of ideally fast closed-loop dynamics of the underlying velocity controllers the system can be described in the form $\dot{\mathbf{x}} = \mathbf{f}(\mathbf{x}, \mathbf{u})$ with the state vector $\mathbf{x}^{\mathrm{T}} = [\mathbf{q}^{\mathrm{T}}, \dot{\mathbf{q}}^{\mathrm{T}}]$ and the accelerations $\ddot{\mathbf{q}}_A$ as the new control input $\mathbf{u} = \dot{\mathbf{v}}$. Neglecting the small damping of the cable's pendulum motion the dynamical system is given by the following set of ordinary differential equations (ODEs)

$$\ddot{s}_1 = u_1 , \quad \ddot{s}_2 = u_2 , \quad \ddot{\phi}_1 = u_3 , \tag{1a}$$

$$\begin{aligned} \ddot{\phi}_2 = \frac{1}{s_2 \cos\phi_3} \Big(& \left(s_2\dot{\phi}_1^2 \sin\phi_2 - 2s_2\dot{\phi}_1\dot{\phi}_3\right)\cos\phi_2\cos\phi_3 - 2\dot{s}_2\dot{\phi}_2\cos\phi_3 - g\sin\phi_2 \\ & + \left(u_1 - s_1\dot{\phi}_1^2 - \left(s_2u_3 + 2\dot{s}_2\dot{\phi}_1\right)\sin\phi_3\right)\cos\phi_2 + 2s_2\dot{\phi}_2\dot{\phi}_3\sin\phi_3 \Big) , \end{aligned} \tag{1b}$$

$$\begin{aligned} \ddot{\phi}_3 = \frac{1}{s_2} \Big(& \left(s_1\dot{\phi}_1^2 - u_1\right)\sin\phi_2\sin\phi_3 + \left(2s_2\dot{\phi}_1\dot{\phi}_2\cos^2\phi_3 - g\sin\phi_3\right)\cos\phi_2 \\ & -2\dot{s}_2\dot{\phi}_3 + \left(s_2u_3 + 2\dot{s}_2\dot{\phi}_1\right)\sin\phi_2 \\ & - \left(s_1u_3 + 2\dot{s}_1\dot{\phi}_1 - \left(s_2\dot{\phi}_1^2\cos^2\phi_2 - s_2\dot{\phi}_2^2\right)\sin\phi_3\right)\cos\phi_3 \Big) \end{aligned} \tag{1c}$$

with the gravitational constant g. It can be seen from (1) that the idealized underlying velocity controllers entail that the system equations are independent of the load mass m_L as well as of friction effects in the actuated axes. Henceforth, the dynamical model (1) will serve as the basis for the controller design in the subsequent section.

3 Combined path planning and control concept

In this section, a combined path planning and control concept is introduced which is based on a user defined set-point change for the global load position within the three-dimensional work space of the crane. A path planning algorithm thus provides a suitable trajectory from the start- to the end-point for the subsequent flatness-based feedforward control by taking into account the information about the constrained work space (e.g. provided by computer vision). In order to track the desired trajectory also in the case of disturbances and model inaccuracies a closed-loop

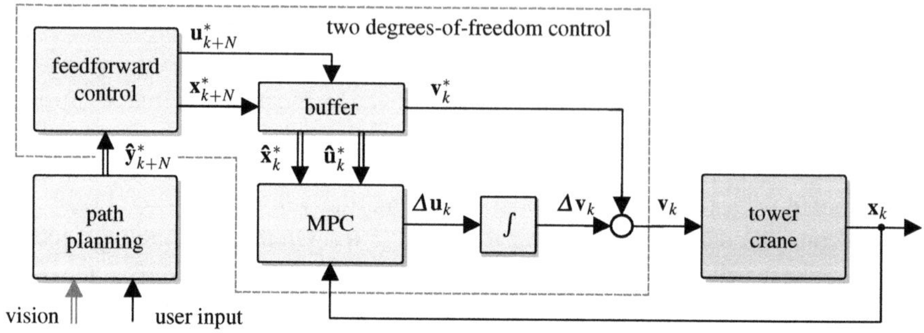

Fig. 2 Two degrees-of-freedom control and path planning concept for the tower crane

model predictive control (MPC) is designed for the trajectory error system, thus constituting a two degrees-of-freedom control structure. Figure 2 illustrates the overall concept which will be described in more detail in the following subsections.

3.1 Flatness-based feedforward control

Following the idea of [3], it can be shown that a so-called flat output of the system (1) can be found in terms of the global load position. In Cartesian coordinates it can be written in the form $\mathbf{y} = [x_L, y_L, z_L]^{\mathrm{T}}$ with

$$x_L = \cos\phi_1\,(s_1 - s_2 \sin\phi_2 \cos\phi_3) - s_2 \sin\phi_1 \sin\phi_3, \tag{2a}$$

$$y_L = \sin\phi_1\,(s_1 - s_2 \sin\phi_2 \cos\phi_3) + s_2 \cos\phi_1 \sin\phi_3, \tag{2b}$$

$$z_L = h - s_2 \cos\phi_2 \cos\phi_3. \tag{2c}$$

By introducing an additional system variable, namely the cable force F_C, the equilibrium of forces at the load can be written as

$$m_L \ddot{x}_L = F_C\,(\cos\phi_1 \sin\phi_2 \cos\phi_3 + \sin\phi_1 \sin\phi_3), \tag{3a}$$

$$m_L \ddot{y}_L = F_C\,(\sin\phi_1 \sin\phi_2 \cos\phi_3 - \cos\phi_1 \sin\phi_3), \tag{3b}$$

$$m_L\,(g + \ddot{z}_L) = F_C \cos\phi_2 \cos\phi_3. \tag{3c}$$

By eliminating s_1, s_2, ϕ_1, ϕ_2, ϕ_3 and F_C from (2) and (3a) all system variables can be parametrized as functions of x_L, y_L, z_L, $\ddot{x}_L$, $\ddot{y}_L$ and $\ddot{z}_L$. Hence, the input velocities $\dot{s}_1$, $\dot{s}_2$ and $\dot{\phi}_1$ can also be calculated as functions of the flat output and its time derivatives yielding the feedforward control law $\mathbf{v}^* = [\dot{s}_1^*, \dot{s}_2^*, \dot{\phi}_1^*]^{\mathrm{T}}$ for a desired trajectory $\mathbf{y}^* = [x_L^*, y_L^*, z_L^*]^{\mathrm{T}}$. The feedback control, as will be outlined in Subsect. 3.2, however, additionally requires the parametrization of the input accelerations $\mathbf{u}^* = \dot{\mathbf{v}}^*$, thus necessitating a four times differentiable trajectory $\mathbf{y}^*$.

In the real-time implementation the desired trajectory $\mathbf{y}^*$ has to be provided by the path planning in the form

$$\hat{\mathbf{y}}_k^* = \left\{ \mathbf{y}_k^*, \dot{\mathbf{y}}_k^*, \ldots, \mathbf{y}_k^{*(r-1)} \right\}$$

with $\mathbf{y}_k^* = \mathbf{y}^*(t_k)$ at each discrete time step $t_k = k\Delta t$, the fixed sampling time Δt and $r = 5$. Furthermore, the desired state vector $\mathbf{x}_k^* = \mathbf{x}^*(t_k)$ and the desired input acceleration $\mathbf{u}_k^* = \mathbf{u}^*(t_k)$ are calculated in the feedforward control block, see Fig. 2.

3.2 *Model predictive feedback control*

In the case of disturbances or model inaccuracies, the feedback controller in Fig. 2 has to stabilize the tracking error $\Delta\mathbf{x}(t) = \mathbf{x}(t) - \mathbf{x}^*(t)$ by determining the control action $\Delta\mathbf{u}(t) = \mathbf{u}(t) - \mathbf{u}^*(t)$ that corrects the feedforward control $\mathbf{u}^*(t)$. Summarizing the ODEs (1) in the form $\dot{\mathbf{x}} = \mathbf{f}(\mathbf{x}, \mathbf{u})$, the tracking error $\Delta\mathbf{x}$ is described by the nonlinear and time-varying error dynamics

$$\Delta\dot{\mathbf{x}}(t) = \mathbf{f}\left(\mathbf{x}^*(t) + \Delta\mathbf{x}(t), \mathbf{u}^*(t) + \Delta\mathbf{u}(t)\right) - \dot{\mathbf{x}}^*(t) =: \mathbf{F}(\Delta\mathbf{x}, \Delta\mathbf{u}, t)\,. \quad (4)$$

The model predictive controller accounts for the error dynamics (4) by solving the following optimal control problem (OCP) in a discrete-time fashion for each sampling instant $t_k = k\Delta t$ with the sampling time Δt:

$$\min_{\Delta\mathbf{u}(\cdot)} \quad J(\Delta\mathbf{u}(\cdot), \Delta\mathbf{x}_k) := \int_{t_k}^{t_k+T} \Delta\mathbf{x}^{\mathrm{T}}(t)\mathbf{Q}\Delta\mathbf{x}(t) + \Delta\mathbf{u}^{\mathrm{T}}(t)\mathbf{R}\Delta\mathbf{u}(t)\,\mathrm{d}t \quad (5a)$$

$$\text{s.t.} \quad \Delta\dot{\mathbf{x}}(t) = \mathbf{F}(\Delta\mathbf{x}(t), \Delta\mathbf{u}(t), t)\,, \quad \Delta\mathbf{x}(t_k) = \Delta\mathbf{x}_k \quad (5b)$$

$$\Delta\mathbf{u}(t) \in [\,\Delta\mathbf{u}^-(t), \Delta\mathbf{u}^+(t)\,]\,, \quad t \in [\,t_k, t_k + T\,] \quad (5c)$$

with the time-varying control input constraints $\Delta\mathbf{u}^{\pm}(t) := \mathbf{u}^{\pm} - \mathbf{u}^*(t)$. Starting from the tracking error $\Delta\mathbf{x}_k$ at time t_k, the error dynamics are used to predict the error trajectory $\Delta\mathbf{x}(t)$ over a finite horizon $t \in [\,t_k, t_k + T\,]$ with the length $T = N\,\Delta t$ and an integer number of N discrete sampling periods. The cost functional (5a) to be minimized penalizes the error $\Delta\mathbf{x}(t)$ as well as the control action $\Delta\mathbf{u}(t)$ w.r.t. the positive semi-definite matrices $\mathbf{Q}$ and $\mathbf{R}$.

The OCP (5) is solved in each time step t_k and $\Delta\mathbf{u}_k := \Delta\mathbf{u}(t_k)$ of the computed trajectory $\Delta\mathbf{u}(t)$ is used as feedback control for $t \in [\,t_k, t_{k+1}\,)$ in Fig. 2. In order to guarantee real-time feasibility for the crane system, an incremental optimization strategy was used that computes a suboptimal solution that is refined over the single MPC steps as proposed in [4]. The optimization algorithm consists of a projected gradient method with an adaptive line search method, see [5, 6].

Solving the OCP (5) at each sampling instant t_k, however, requires the desired trajectories of the state and input vectors $\hat{\mathbf{x}}^*$ and $\hat{\mathbf{u}}^*$ over the complete finite horizon of the length T. Hence, the path planning and feedforward control have to calculate the desired trajectories N discrete time steps in advance such that the sets

$$\hat{\mathbf{x}}_k^* = \{\mathbf{x}_k^*, \mathbf{x}_{k+1}^*, \ldots, \mathbf{x}_{k+N}^*\}, \quad \hat{\mathbf{u}}_k^* = \{\mathbf{u}_k^*, \mathbf{u}_{k+1}^*, \ldots, \mathbf{u}_{k+N}^*\}$$

can be provided by a suitable ring buffer, cf. Fig. 2. Moreover, note that the actual input velocities $\Delta\mathbf{v}_k$ are generated by a time discrete integrator, while the desired input velocities $\mathbf{v}_k^*$ can be taken directly from the buffered state vector $\mathbf{x}_k^*$, see Fig. 2.

3.3 Path planning

Upon discussing the two degrees-of-freedom control concept based on suitable trajectories $\mathbf{y}^*$, in a next step the actual path planning is considered accounting for arbitrary obstacles located within the work space of the crane. For this, a two-stage concept has been chosen, where an approximated path is calculated for a discretized work space before a sufficiently smooth trajectory is planned along the discrete path.

In a first step, the work space of the crane is discretized into a finite number of hollow-cylindrical bricks. Therefore, it is advantageous to introduce a transformation from Cartesian to cylindrical coordinates of the form $(x_L, y_L, z_L) \mapsto (r_L, \phi_L, z_L)$ with $x_L = r_L \cos\phi_L$ and $y_L = r_L \sin\phi_L$. Assuming that the information of possible obstacles in the work space is provided (e.g. by a 3d-capable camera system) in terms of marked (not permitted) finite work space elements, the so-called A^* search algorithm is utilized to calculate a discrete admissible path from the start- to a predefined end-point. The A^* algorithm, see, e.g., [7], is a widespread tool in particular for graph traversal and path finding. Here, the center points of the finite work space elements are utilized as the discrete nodes with the 2-norm between two neighboring nodes in cylindrical coordinates serving as the path-cost function and the corresponding 1-norm serving as the so-called heuristic estimate. If an admissible path can be found the A^* algorithm provides the path with the smallest possible distance in terms of a set of neighboring nodes from the start- to the end-point.

Finally, a sufficiently smooth desired trajectory $\mathbf{y}^*(\tau)$ is computed from the set of discrete nodes in terms of a three-dimensional B-spline curve of degree $r = 5$, see [8,9]. With a suitable time parametrization of the curve parameter $\tau(t)$ the trajectory $\mathbf{y}_k^*$ is provided to the feedforward control at each sampling instant t_k, cf. Fig. 2.

4 Measurement results

An experimental validation of the proposed path planning and control concepts has been performed at the Automation and Control Institute, Vienna University of Technology (see http://www.acin.tuwien.ac.at/fileadmin/craneV1.wmv) by utilizing

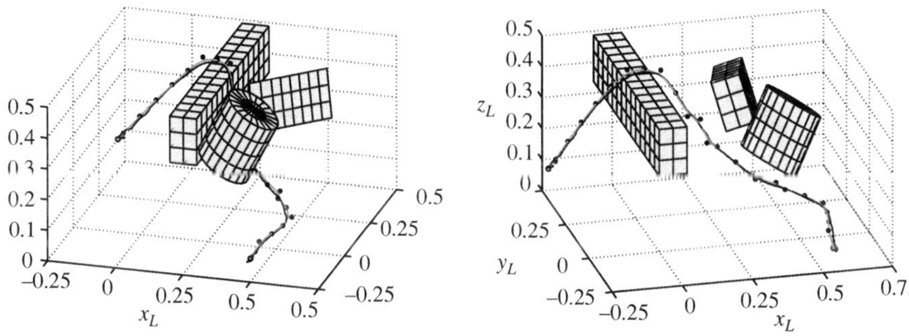

Fig. 3 Discrete path (*), desired trajectory (*gray*) and measured trajectory (*black*)

a laboratory model of the tower crane with a height of $h = 1$ m and a work space of $r_L \in [0.15, 0.8]$ m, $\phi_L \in [-150, 150]^\circ$ and $z_L \in [0, 0.8]$ m. The system is actuated by three velocity controlled dc-motors as described in Section 1. The five DOFs s_1, s_2, ϕ_1, ϕ_2 and ϕ_3 are measured with incremental encoders while the respective velocities are estimated using approximate differentiation.

The path planning algorithm as well as the feedforward and feedback controllers are implemented in the real-time operating system Matlab xPC Target on an Intel CoreDuo CPU. Thereby, the sampling time of the controllers is chosen as $\Delta t = 2$ ms with a moving horizon period for the MPC of $T = 1$ s. The discrete path planning algorithm is operating with a sampling rate of $\Delta t_P = 50$ ms. However, since the calculation time and thus the required number of iterations for the path planning strongly depends on the current work space scenario the maximum number of iterations is fixed and in case of a CPU overload the intermediate result is carried over to the next calculation step. The chosen work space discretization added up to 16.000 elements.

Now, a test scenario with three obstacles (two cuboids and one cylinder) is considered together with a desired set-point change from the start-point r_L=0.5 m, ϕ_L=120° and z_L=0.15 m to the end-point r_L=0.6 m, $\phi_L = -30°$ and z_L=0.15 m. Figure 3 illustrates the work space and the 3d-curve from the start- to the end-point in terms of the discrete path as well as the desired and measured trajectory from two different perspectives. Thereby, no external disturbances were acting on the system while the same scenario with an impulse-type disturbance applied to the load is depicted in Fig. 4. The corresponding Figs. 5 and 6 illustrate the respective time evolutions of the systems variables, i.e. the DOFs s_1, s_2, ϕ_1, ϕ_2 and ϕ_3 as well as the input velocities $\dot{s}_1$, $\dot{s}_2$ and $\dot{\phi}_1$. As can be observed from Fig. 5 (in particular from the time evolution of the angle ϕ_3) the impulse-type disturbance is occurring at the time $t \simeq 5$ s. The thus resulting error is efficiently corrected by the feedback controller by means of the input velocities, cf. Fig. 6 (see in particular the deviation of the angular velocity $\dot{\phi}_1$). The residual offsets of the angles ϕ_2 and ϕ_3 as well as the non vanishing input velocities $\dot{s}_1$, $\dot{s}_2$ and $\dot{\phi}_1$ in steady state are due to stick-slip effects at the gimbal and the actuated axes (gear boxes, etc.), respectively.

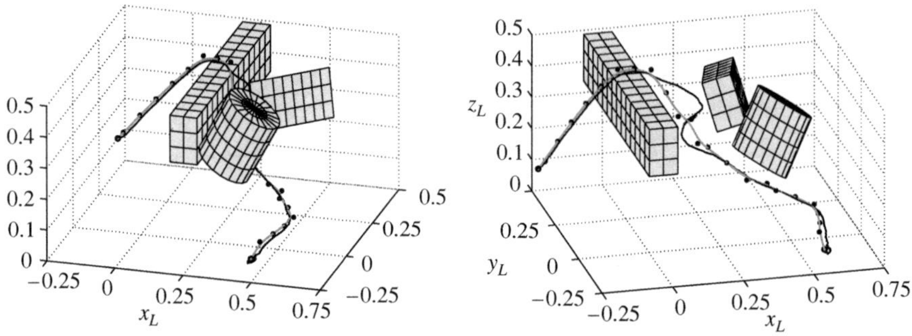

Fig. 4 Discrete path (*), desired trajectory (*gray*) and measured trajectory (*black*) with disturbance

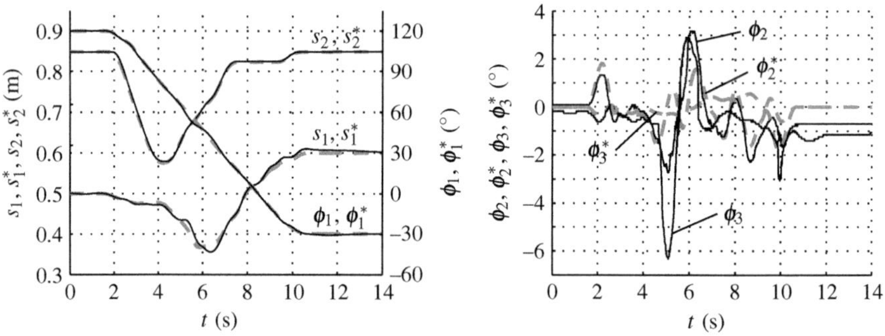

Fig. 5 Degrees-of-freedom, time evolution (*dashed line* desired, *solid line* measured quantities)

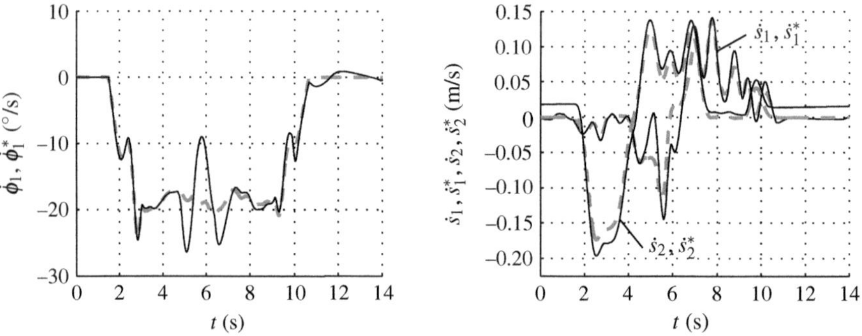

Fig. 6 Input quantities, time evolution (*dashed line* desired, *solid line* measured quantities)

The above presented measurement results reveal the advantages of the proposed concept where the path planning and flatness-based feedforward control allow for an efficient transition between the start- and end-point, while at the same time the MPC guarantees a good performance in terms of stabilization and disturbance rejection. Current research is spent on the integration of a smart camera network for workspace supervision as well as on the development of an online trajectory re-planning scheme in order to cope with frequently updated camera information and moving obstacles within the work space.

Acknowledgements This work was partially supported by the Austrian FIT–IT project CLIC under project no. 819482 as well as by the Austrian FWF project no. P21253-N22 "Constrained Trajectory Optimization".

References

1. Huang, W.H., Ahuja, N.: Gross motion planning – a survey. ACM Comput. Surv. **24**, 219–291 (1992)
2. Jia, Z., Balasuriya, A., Challa, S.: Recent developments in vision based target tracking for autonomous vehicles navigation. In: IEEE Intelligent Transportation Systems Conference, pp. 765–770. Toronto, Canada (2006)
3. Fliess, M., Lévine, J., Martin, P., Rouchon, P.: Flatness and defect of non-linear systems: introductory theory and examples. Int. J. Contr. **61**, 1327–1361 (1995)
4. Graichen, K., Kugi, A.: Stability and incremental improvement of suboptimal MPC without terminal constraints. IEEE Trans. Automat. Contr. **55**(11), 2576–2580 (2010)
5. Graichen, K., Egretzberger, M., Kugi, A.: Suboptimal model predictivecontrol of a laboratory crane. 8th IFAC Symposium on Nonlinear Control Systems, pp. 397–402. Bologna, Italy (2010)
6. Graichen, K., Egretzberger, M., Kugi, A.: Ein suboptimaler Ansatz zur schnellen modell-prädiktiven Regelung nichtlinearer Systeme, vol. 58, pp. 447–456. at-Automatisierungstechnik, Oldenbourg Verlag (2010)
7. Hart, P.E., Nilsson, N.J., Raphael, B.: A formal basis for the heuristic determination of minimum cost paths. IEEE Trans. Syst. Sci. Cybernet. **4**, 100–107 (1968)
8. LaValle, S.M.: Planning Algorithms. Cambridge University Press, USA (2006)
9. LaTombe, J.-C.: Robot Motion Planning. Kluwer Academic Publishers, Boston, USA (2004)

Stress Focusing Effect of an Elastic Solid in the Context of Generalized Thermoelasticity

T. Furukawa

Abstract The stress focusing effect in a solid sphere and an infinitely long solid cylinder under instantaneous uniform heating at the free surface is studied on the basis of the generalized thermoelastic theories, that is, Lord-Shulman and Green-Lindsay theories. The combined governing equations of both theories are solved by Laplace transform. Calculations have been performed to exhibit the radial distributions on the basis of Lord-Shulman theory. The effects of thermomechanical coupling parameter and relaxation time on the stress focusing phenomena as well as the singularity of stresses are discussed.

1 Introduction

The theory of dynamic thermoelasticity, which takes into account the coupling effects between temperature and strain fields, involves the infinite thermal wave speed. The theory of generalized thermoelasticity has been developed in an attempt to eliminate the physical paradox of the infinite velocity of thermal wave. At present, there are two theories of generalized thermoelasticity: the first is proposed by Lord and Shulman [1], the second is proposed by Green and Lindsay [2]. There are a lot of papers in the context of these theories, for example, Lord-Shulman [3,4] and Green-Lindsay [5, 6]. Recently, other generalized theories have presented (see Ignaczak and Hetnarski [7] and Ignaczak and Ostoja-Starzewski [8]), for example, Green-Nagdhi theory [9]. Noda, Furukawa, and Ashida used the fundamental equations of generalized thermoelasticity introduced by us [10], which include Lord-Shulman theory and Green-Lindsay theory, and analyzed several problems.

T. Furukawa (✉)
Department of Mechanical Systems Engineering, University of the Ryukyus, Senbaru 1, Nishihara, Okinawa 903-0213, Japan
e-mail: furukawa@teada.tec.u-ryukyu.ac.jp

H. Irschik et al. (eds.), *Advanced Dynamics and Model-Based Control of Structures and Machines*, DOI 10.1007/978-3-7091-0797-3_9,

The analysis of an infinitely long cylindrical rod subjected to a sudden rise in temperature uniformly over its cross section has been studied by Ho [11]. Due to the instantaneous heating, the stress waves reflected from the cylindrical surface of the rod may accumulate at the center and give rise to very high stresses, even though the initial thermal stress is relatively small. This phenomenon is called the stress focusing effect. Hata [12–14] has studied this effect for several cases of cylindrical rod. The stress focusing effect for spheres has been studied by Mann-Nachbar [15] and Hata [16–19]. However, these series of papers have been studied in the context of classical uncoupled theory of dynamic thermoelasticity, and the condition of temperature is very limited that the uniform temperature change all over the sphere or cylinder is occurred at the initial time. In this paper, we treat an isotropic and homogeneous solid sphere and infinitely long solid cylinder. We use the fundamental equations of generalized thermoelasticity, which include two theories, that is, Lord-Shulman theory and Green-Lindsay theory.

2 Analysis

Fundamental equations combined two generalized theories, that is, Lord-Shulman theory and Green-Lindsay theory, are expressed as following five equations.

Strain-displacement relation

$$\varepsilon_{ij} = \frac{1}{2}\left(u_{i,j} + u_{j,i}\right), \tag{1}$$

Equation of motion

$$\sigma_{ij},_j + F_i = \bar{\rho} u_i,_{tt}, \tag{2}$$

Energy equation

$$-q_{ti},_i + W = \bar{\rho}\, c_v (T + \delta_{2k} t_0 T,_t),_t + (3\lambda + 2\mu)\alpha T_0 \varepsilon_{kk},_t, \tag{3}$$

Stress-strain-temperature relation

$$\sigma_{ij} = 2\mu\varepsilon_{ij} + \lambda\varepsilon_{kk}\delta_{ij} - (3\lambda + 2\mu)\alpha(T + \delta_{2k} t_1 T,_t)\delta_{ij}, \tag{4}$$

Heat equation [Modified Fourier's law]

$$q_{ti} + \delta_{1k} t_0 q_{ti},_t = -\bar{k} T,_i . \tag{5}$$

Here, ε_{ij}: strain component, σ_{ij}: stress component, u_i: displacement component, F_i: body force component, $\bar{\rho}$: density, t: time, q_t: heat flux, W: heat supply, c_v: specific heat at constant volume, T: temperature difference from initial state absolute temperature T_0, α: coefficient of linear thermal expansion, λ and μ: Lame's constants, ε_{kk} : volumetric strain, $\bar{k}$: thermal conductivity, t_0 and t_1: relaxation times, and δ_{jk}: Kronecker's delta whose subscript k denotes the number of relaxation times. When we put $k = 0, 1, 2$, (3)–(5) are coincided to classical, Lord-Shulman, and Green-Lindsay theories, respectively. The comma denotes the differentiation with following variable.

We consider a solid sphere and an infinitely long solid cylinder under the sudden rise in temperature at the free surface at time $t = 0$ and we ignore body force and internal heat supply. In this case, the fundamental equations consist of three equations, that is, the heat conduction equation combined with (3)–(5)

$$\kappa \nabla^2 T - (T + t_0 T_{,t})_{,t} = \frac{\delta}{\zeta\alpha}(e + \delta_{1k} t_0 e_{,t})_{,t}, \tag{6}$$

the equation of motion represented by displacement component

$$u_{,rr} + \frac{n}{r}\left(u_{,r} - \frac{u}{r}\right) - \zeta\alpha\,(T + \delta_{2k} t_1 T_{,t})_{,r} = \frac{1}{v_e^2} u_{,tt}, \tag{7}$$

and the stress–strain-temperature relations

$$\begin{bmatrix} \sigma_{rr} \\ \sigma_{\theta\theta} \end{bmatrix} = 2\mu \begin{bmatrix} u_{,r} \\ \frac{u}{r} \end{bmatrix} + \lambda\left(u_{,r} + \frac{n}{r}u\right) - (3\lambda + 2\mu)\alpha\,(T + \delta_{2k} t_1 T_{,t}), \tag{8}$$

where

$$\nabla^2 = \frac{\partial^2}{\partial r^2} + \frac{n}{r}\frac{\partial}{\partial r}, \quad e = u_{,r} + \frac{n}{r}u \quad \text{and} \quad n = \begin{cases} 1: \text{ cylinder} \\ 2: \text{ sphere} \end{cases}$$

Here, σ_{rr} and $\sigma_{\theta\theta}$: radial stress and hoop stress, respectively, κ: thermal diffusivity, δ: thermomechanical coupling parameter, v_e : velocity of longitudinal wave.

We introduce the dimensionless quantities:

$$\rho = \frac{r}{a}, \quad \beta = \frac{\kappa}{v_e a}, \quad (\tau, \tau_0, \tau_1) = \frac{v_e}{a}(t, t_0, t_1), \quad \theta = \frac{T}{T_1},$$

$$U = \frac{u}{\zeta\alpha T_1 a}, \quad (\bar{\sigma}_{rr}, \bar{\sigma}_{\theta\theta}) = \frac{(\sigma_{rr}, \sigma_{\theta\theta})}{(3\lambda + 2\mu)\alpha T_1}, \tag{9}$$

where T_1 is a prescribed temperature. Substituting (9) into (6)–(8) and applying Laplace transform under the initial condition

$$\tau = 0: \; U = U_{,\tau} = 0, \; \theta = \theta_{,\tau} = 0, \tag{10}$$

we have

$$\bar{\nabla}^2\theta^* - \frac{1}{\beta}s(1+\tau_0 s)\theta^* = \frac{\delta}{\beta}s(1+\delta_{1k}\tau_0 s)\left(U^*{}_{,\rho} + \frac{n}{\rho}U^*\right), \tag{11}$$

$$\bar{\nabla}^2 U^* - \left(s^2 + \frac{n}{\rho^2}\right)U^* = (1+\delta_{2k}\tau_1 s)\theta^*{}_{,\rho}, \tag{12}$$

$$\begin{bmatrix} \overline{\sigma}^*_{rr} \\ \overline{\sigma}^*_{\theta\theta} \end{bmatrix} = \begin{bmatrix} 1 \\ \eta \end{bmatrix}\left(U^*{}_{,\rho} + \frac{n}{\rho}U^*\right) + \begin{bmatrix} n \\ -1 \end{bmatrix}(\eta - 1)\frac{U^*}{\rho} - (1+\delta_{2k}\tau_1 s)\theta^*, \tag{13}$$

where the asterisk (*) denotes the Laplace transform, and s is Laplace transform parameter.

From (11) and (12), we obtain the equation for displacement

$$\left(\bar{\nabla}^2 - \frac{n}{\rho^2} - k_1^2\right)\left(\bar{\nabla}^2 - \frac{n}{\rho^2} - k_2^2\right)U^* = 0, \tag{14}$$

where k_1 and k_2 are positive roots of the following characteristic equation

$$k^4 - s(B_0 s + B_1)k^2 + \frac{1}{\beta}(1+\tau_0 s)s^3 = 0, \tag{15}$$

with

$$B_0 = 1 + \frac{1}{\beta}[\tau_0 + \delta(\delta_{1k}\tau_0 + \delta_{2k}\tau_1)], \quad B_1 = \frac{1+\delta}{\beta}. \tag{16}$$

From (14), the radial displacement U^*can be split as follows

$$U^* = U_1^* + U_2^*, \tag{17}$$

where

$$\left(\bar{\nabla}^2 - \frac{n}{\rho^2} - k_i^2\right)U_i^* = 0. \ (i = 1,\ 2) \tag{18}$$

Firstly we obtain the solutions for infinitely long solid cylinder. The solution of the differential equation (18) is

$$U_i^* = \zeta(1+\delta_{2k}\tau_1 s)A_i k_i I_1(k_i\rho), \tag{19}$$

where $I_1()$ is the modified Bessel function of first kind with first order and A_i is unknown coefficient. Therefore the radial displacement is expressed as follows

$$U^* = \zeta(1+\delta_{2k}\tau_1 s)[A_1 k_1 I_1(k_1\rho) + A_2 k_2 I_1(k_2\rho)]. \tag{20}$$

Similarly, the temperature and stresses are shown in

$$\theta^* = \left(k_1^2 - s^2\right) A_1 I_0\left(k_1\rho\right) + \left(k_2^2 - s^2\right) A_2 I_0\left(k_2\rho\right), \tag{21}$$

$$\begin{bmatrix} \bar{\sigma}_{rr}^* \\ \bar{\sigma}_{\theta\theta}^* \end{bmatrix} = \zeta(1+\delta_{2k}\tau_1 s)\begin{bmatrix} 1 \\ \eta \end{bmatrix}\left[A_1 k_1^2 I_0\left(k_1\rho\right) + A_2 k_2^2 I_0\left(k_2\rho\right)\right]$$
$$-\begin{bmatrix} 1 \\ -1 \end{bmatrix}(1-\eta)\frac{U^*}{\rho} - (1+\delta_{2k}\tau_1 s)\,\theta^*, \tag{22}$$

where $I_0(\ )$ is the modified Bessel function of first kind with 0th order.

Secondly we obtain the solutions for solid sphere. The solution of the differential equation (18) is

$$U_i^* = -\zeta(1+\delta_{2k}\tau_1 s)A_i\frac{1}{\rho^2}(\sinh k_i\rho - k_i\rho\cosh k_i\rho). \tag{23}$$

Therefore the radial displacement is expressed as follows

$$U^* = -\zeta(1+\delta_{2k}\tau_1 s)\frac{1}{\rho^2}[A_1(\sinh k_1\rho - k_1\rho\cosh k_1\rho)$$
$$+ A_2(\sinh k_2\rho - k_2\rho\cosh k_2\rho)]. \tag{24}$$

Similarly, the temperature and stresses are shown in

$$\theta^* = (k_1^2 - s^2)A_1\frac{1}{\rho^2}\sinh k_1\rho + (k_2^2 - s^2)A_2\frac{1}{\rho^2}\sinh k_2\rho, \tag{25}$$

$$\bar{\sigma}_{rr}^* = \zeta(1+\delta_{2k}\tau_1 s)\sum_{i=1}^{2}\frac{A_i}{\rho}\left[s^2\sinh k_i\rho + 2\frac{1-\eta}{\rho^2}(\sinh k_i\rho - k_i\rho\cosh k_i\rho)\right],$$

$$\bar{\sigma}_{\theta\theta}^* = \zeta(1+\delta_{2k}\tau_1 s)\sum_{i=1}^{2}\frac{A_i}{\rho}\left[\left\{k_i^2(\eta-1)+s^2\right\}\sinh k_i\rho\right.$$
$$\left. -\frac{1-\eta}{\rho^2}(\sinh k_i\rho - k_i\rho\cosh k_i\rho)\right]. \tag{26}$$

Next we determine the unknowns A_1 and A_2. The dimensionless boundary conditions of Laplace transformed domain are

$$\rho = 1: \quad \theta^* = \frac{1}{s}, \quad \bar{\sigma}_{rr}^* = 0. \tag{27}$$

Introduction (21) and (22a) into (27) leads the coefficients A_1 and A_2 for solid cylinder as follows.

$$A_1 = \frac{s^2 I_0(k_2) - (1-\eta)k_2 I_1(k_2)}{(c_1 - c_2)s},$$
$$A_2 = -\frac{s^2 I_0(k_1) - (1-\eta)k_1 I_1(k_1)}{(c_1 - c_2)s}, \tag{28}$$

where

$$c_1 = (k_1^2 - s^2)I_0(k_1)[k_2^2 I_0(k_2) - (1-\eta)k_2 I_1(k_2)],$$
$$c_2 = (k_2^2 - s^2)I_0(k_2)[k_1^2 I_0(k_1) - (1-\eta)k_1 I_1(k_1)]. \tag{29}$$

Introduction (25) and (26a) into (27) leads the coefficients A_1 and A_2 for solid sphere as follows.

$$A_1 = \frac{s^2 \sinh k_2 + 2(1-\eta)(\sinh k_2 - k_2 \cosh k_2)}{cs},$$
$$A_2 = -\frac{s^2 \sinh k_1 + 2(1-\eta)(\sinh k_1 - k_1 \cosh k_1)}{cs}, \tag{30}$$

where

$$\begin{aligned} c = {} & (k_1^2 - k_2^2)s^2 \sinh k_1 \sinh k_2 \\ & + 2(1-\eta)\left[(k_1^2 - s^2)\sinh k_1(\sinh k_2 - k_2\cosh k_2)\right. \\ & \left. -(k_2^2 - s^2)\sinh k_2(\sinh k_1 - k_1 \cosh k_1)\right]. \end{aligned} \tag{31}$$

So that, the solutions for temperature, displacements and thermal stresses are obtained in the Laplace transformed domains. Because the analytical inversions are much difficult, so the inversions are carried out numerically.

3 Numerical results and discussion

We calculate for Poisson's ratio $\nu = 0.3$ and the inertia parameter $\beta = 0.1$ in the context of Lord-Shulman theory. Firstly, we show the figures for solid sphere. Figure 1 shows the radial stress distribution at the several times. To have a clear view, only the curves of stress waves propagating in one direction are displayed here. In Fig. 1a, the positions of peaks correspond to the positions of the stress wave fronts at different instants. In both classical and generalized cases, with the stress wave fronts approaching the center, the peaks of the stresses increase at a

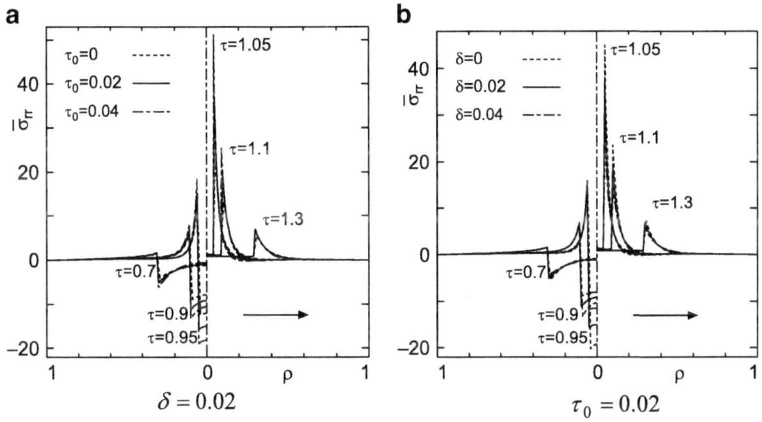

Fig. 1 The radial stress distributions of solid sphere

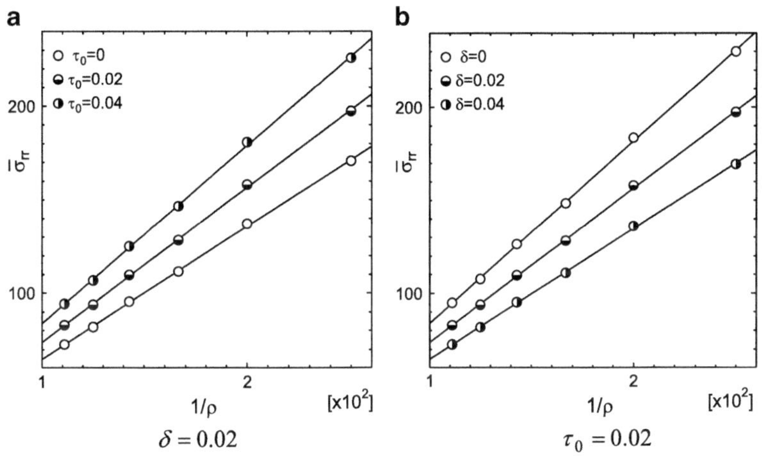

Fig. 2 Curves of $\sigma_{rr} - 1/\rho$ near the center of sphere

great rate. This phenomenon shows the stress-focusing effect. Comparing the results for different relaxation times shows that the generalized theory predicts larger peaks of stress waves than the classical theory, and with an increase of the relaxation time, the peaks increase. In Fig. 1b, the behavior of the stress wave is shown for a fixed relaxation time. It can be seen that the stress-focusing effect occurs for both coupled and uncoupled cases. The comparison of the results of different coupling parameters shows that increasing the thermomechanical coupling parameter results in a decrease of the peak. Figure 2 shows the relation of radial stress and reciprocal of the radius. From this figure, it is found that the stresses are proportional to $1/\rho$. This figure indicates the singularity of stresses at the center of the sphere. For the

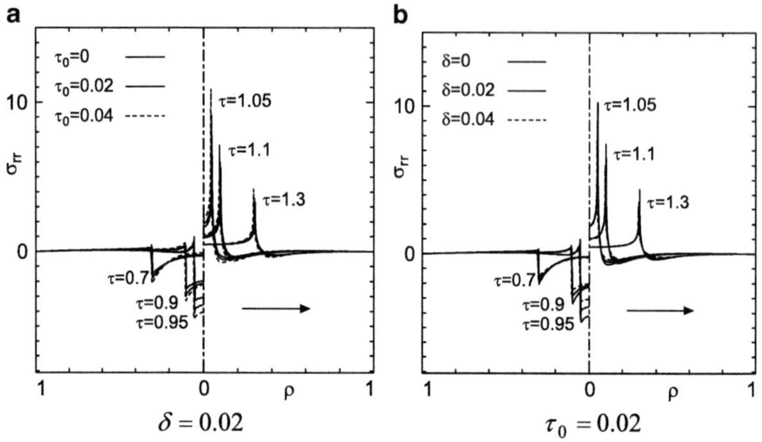

Fig. 3 The radial stress distributions of solid cylinder

classical case, Hata showed that the order of singularity of the stresses is $1/\rho^3$ when he studied a solid sphere subjected to a sudden rise in temperature uniformly over its cross section. The present result disagrees with that of Hata.

Secondly, we calculate for infinitely long solid cylinder in the context of Lord-Shulman theory. Figure 3 shows the radial stress distribution. From these figures, the tendency is very similar to that for sphere, but the stress values are relatively small.

Previously, Ding, Furukawa, and Nakanishi [20] showed that the stress singularity of stresses for solid cylinder was $1/\rho^2$ same as Hata. It seems that the stress singularity of the sphere is larger than that of the cylinder because the stress is concentrated easily on the sphere compared with the cylinder. Therefore, a further examination is necessary to obtain ultimate conclusions.

4 Conclusion

We treated an isotropic and homogeneous solid sphere and infinitely long solid cylinder under instantaneous uniform heating on the free surface at initial time. The fundamental equations of generalized thermoelasticity combined two theories, that is, Lord-Shulman theory and Green-Lindsay theory were used and the exact solutions in Laplace transformed domain could be obtained. Numerical calculations were performed by use of numerical inversion technique. The major results were follows:

1. The stress-focusing effect is occurred for sphere and cylinder under this condition.

2. The stress singularity of generalized theory is larger than that of classical one.
3. The further consideration is needed for determination of the order of stress singularity.

References

1. Lord, H.W., Shulman, Y.: A generalized dynamical theory of thermoelasticity. J. Mech. Phys. Solids **15**, 299–309 (1967)
2. Green, A.E., Lindsay, K.A.: Thermoelasticity. J. Elast. **2**, 1–7 (1972)
3. Youssef, H.M.: Dependence of modulus of elasticity and thermal conductivity on reference temperature in generalized thermoelasticity for an infinite material with a spherical cavity. Appl. Math. Mech. **26**, 470–475 (2005)
4. Sherief, H.H., El-Maghraby, N.M.: A thick plate problem in the theory of generalized thermoelastic diffusion. Int. J. Thermophys. **30**, 2044–2057 (2009)
5. Ghosh, M.K., Kanoria, M.: Generalized thermoelastic functionally graded spherically isotropic solid containing a spherical cavity under thermal shock. Appl. Math. Mech. **29**, 1263–1278 (2008)
6. El-Karamany, A.S., Ezzat, M.A.: Uniqueness and reciprocal theorems in linear micropolar electro-magnetic thermoelasticity with two relaxation times. Mech. Time-Dependent Mater. **13**, 93–115 (2009)
7. Hetnarski, R.B., Ignaczak, J.: Generalized thermoelasticity. J. Therm. Stresses **22**, 451–476 (1999)
8. Ignaczak, J., Ostoja-Starzewski, M.: Thermoelasticity with Finite Wave Speeds. Oxford University Press, London (2009)
9. Yu, J.-G., Xue, T.-L.: Generalized thermoelastic waves in spherical curved plates without energy dissipation. Acta Mech. **212**, 39–50 (2010)
10. Noda, N., Furukawa, T., Ashida, F.: Generalized thermoelasticity in an infinite solid with a hole. J. Therm. Stresses **12**, 385–402 (1989)
11. Ho, C.H.: Stress-focusing effect in a uniformly heated cylindrical rod. Trans. ASME J. Appl. Mech. **43**, 464–468 (1976)
12. Hata, T.: Reconsideration of the stress-focusing effect in a uniformly heated solid cylinder. Trans. ASME J. Appl. Mech. **61**, 676–680 (1994)
13. Hata, T.: Thermal stress-focusing effect following rapid uniform heating of spheres and long cylindrical rods. J. Therm. Stresses **20**, 819–852 (1997)
14. Hata, T., Sumi, N.: Thermal stress-focusing effect in a cylinder with phase transformation. Acta Mech. **195**, 69–80 (2008)
15. Mann-Nachbar, P., Nachbar, W.: Report TR-0066. Aerospace Corp Report TR-0066, pp. 1–49 (1970)
16. Hata, T.: Stress-focusing effect in a solid sphere caused by instantaneous uniform heating. JSME Int. J. I **33**, 33–36 (1990)
17. Hata, T.: Stress-focusing effect in a uniformly heated solid cylinder. Trans ASME J. Appl. Mech. **58**, 58–63 (1991)
18. Hata, T.: Stress-focusing effect in a uniformly heated transversely isotropic sphere. Int. J. Solids Struct. **30**, 1419–1428 (1993)
19. Hata, T.: Thermal stress-focusing effect in a solid sphere subjected to instantaneous heating. J. Therm. Stresses **18**, 437–448 (1995)
20. Ding, X.F., Furukawa, T., Nakanishi, H.: Analysis of the stress-focusing effect in a solid cylinder subjected to instantaneous heating based on the generalized thermoelasticity. J. Therm. Stresses **24**, 383–394 (2001)

Magnetoelastic Effect as Applied to Estimating Elastic-Plastic Strains in Steels and Optimization of Technological Processes

E.S. Gorkunov

Abstract The paper deals with different methods for estimating macro- and microscopic stresses in ferromagnetic materials with the application of magnetic characteristics. It demonstrates the possibility of reconstructing deformation curves by measured magnetic characteristics of steel products.

1 Introduction

Magnetic methods are successfully used to estimate the mechanical properties, structural state and phase composition of steel and cast-iron products. Besides the structural condition, of great importance for structural members is their stress state. Macrostresses and induced anisotropy (texture) can be easily revealed by magnetic methods. To estimate the level of microstresses is a challenge. The estimation methods are based on the recording of parameters associated with the displacement of 90° domain walls, that is, with the redistribution of magnetoelastic energy in a ferromagnet.

Under magnetization reversal, the magnetization of ferromagnets changes discretely in the form of irreversible magnetization jumps of different amplitude (Barkhausen effect). The appearance of Barkhausen jumps (BJs) is due to the interaction of domain walls (DWs) with defects of different nature (pores, inclusions, secondary phase precipitations, dislocations, grain boundaries, stress gradients etc.). Therefore the recorded BJ flux parameters represent the structural condition of ferromagnets, sensitively indicate its changes and can be used both for nondestructive testing and for analysing the degradation of product properties under operation.

E.S. Gorkunov (✉)
Institute of Engineering Science, RAS (Ural Branch), 34 Komsomolskaya str., 620049 Ekaterinburg, Russia
e-mail: ges@imach.uran.ru

H. Irschik et al. (eds.), *Advanced Dynamics and Model-Based Control of Structures and Machines*, DOI 10.1007/978-3-7091-0797-3_10,

2 Estimation of working macrostresses

Iron-based ferromagnets are characterised by the existence of two subsystems of 180° and 90° domain walls (DWs). Figure 1 shows a transition of the 180° domain wall through an inclusion defect in the longitudinal magnetization of an iron-silicon alloy crystal. DW separation from a defect (at a critical field H_{cr}) generates electromagnetic and elastic waves (Fig. 1b).

A similar process takes place in crystal portions with internal residual stresses. The approach of DWs to these portions causes a jumpwise rearrangement of the complex of 90° and 180° domains corresponding to the compression and tension zones [1].

For instance, applied stresses σ_0 change magnetic anisotropy thus reducing the volume of 90° magnetic phases (Fig. 2a, b). At small tensile strains, 90° DWs are substituted by 180° ones, therefore a texture with magnetization vectors lying along the tetragonal axes closest to the direction of tension is formed in the crystals. The area of 180° DWs grows and the processes of magnetization reversal become easier.

Thus, in steels with positive magnetostriction ($\lambda > 0$) under tension ($\sigma > 0$) (positive magnetoelastic effect, $\lambda\sigma > 0$) the magnetization reversal processes become easier, whereas under compression (negative magnetoelastic effect, $\lambda\sigma > 0$)

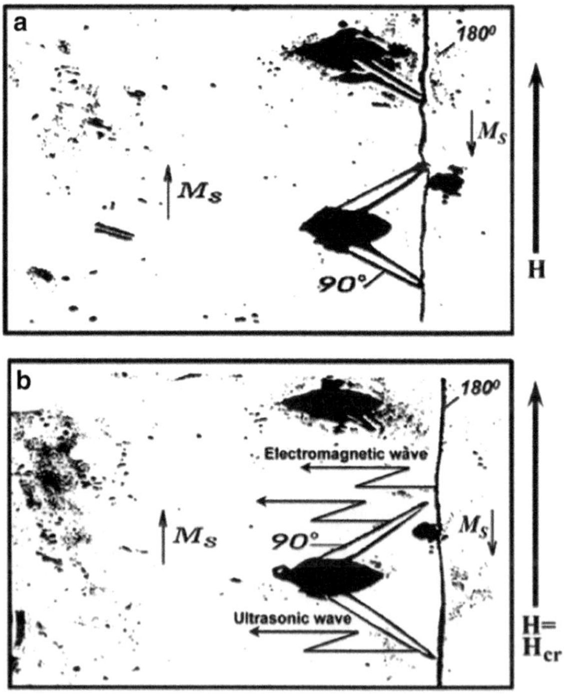

Fig. 1 Domain wall displacement near a defect

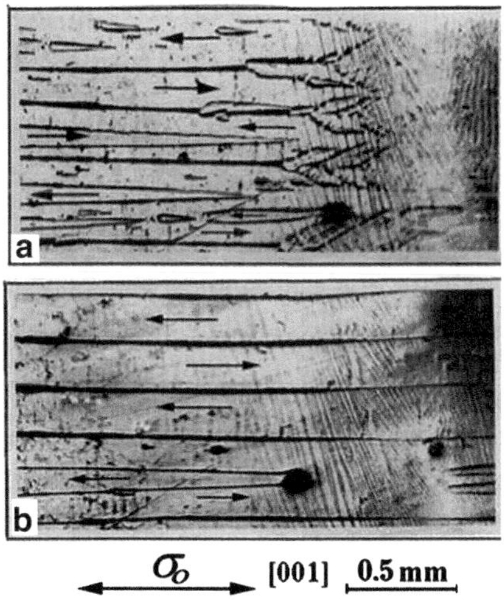

Fig. 2 Rearrangement of 90° domains into 180° ones in silicon iron (**a**) under 100 MPa tension (**b**)

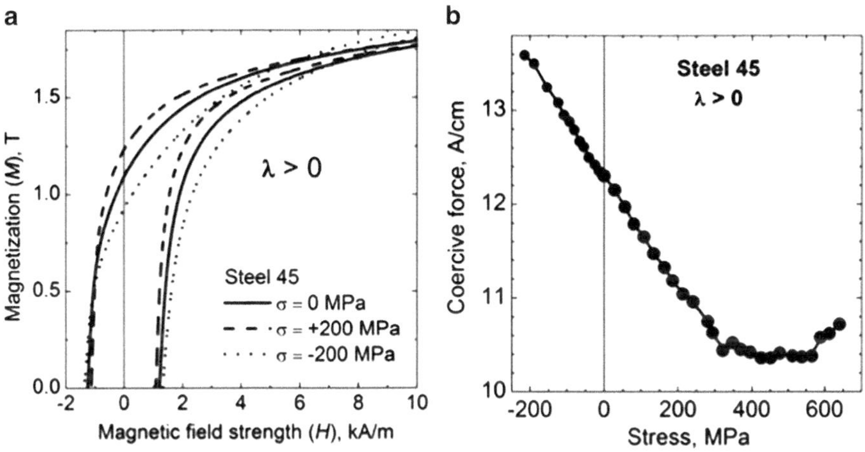

Fig. 3 Hysteresis half-loops for steel 45 (C ~ 0.45%) and coercive force affected by tensile and compressive stresses

they are impeded, and this can be illustrated by magnetic hysteresis loops (Fig. 3a) for steel 45. This principle is a basis for estimating working stresses in members made of ferromagnetic materials (Fig. 3b).

3 Estimation of the level of microstresses

To estimate the microstress level in ferromagnetic materials, three approaches are proposed, which involve magnetic energy redistribution in magnetization reversal during the motion of 90° DWs.

One approach employs creating a predominant displacement of 90° DWs, with the working stresses σ_0 being somewhat equal to internal ones σ_i. When $\sigma_0 \sim \sigma_i$, the magnetic permeability μ^{90} associated with the displacement of 90° walls is much greater than that ensuing from the mobility of 180° ones ($\mu^{90} \gg \mu^{180}$). Since the mobility of 90° DWs is governed by the microstress level, the separated contribution of 90° DWs to the process of magnetization reversal enables the microstress level to be estimated.

Magnetoelastic studies were conducted as follows, Fig. 4: a demagnetized specimen was subjected to tensile stressing (σ_0), then a magnetic field (H) creating induction (B_0) was applied along the tension axis, whereupon low-frequency elastic dynamic vibration of small amplitude ($\Delta\sigma^{\sim}$) was excited in the specimen.

The amplitude of magnetic induction ($B^{\Delta\sigma^{\sim}}$) generated in the specimen by dynamic vibration was recorded by an enveloping coil. The measurements were repeated at different tensile loads ranging between 0 and $0.6\sigma_{0.2}$. The plots of measured induction as dependent on tensile loading (Fig. 5) have a characteristic form with a peak [2, 3].

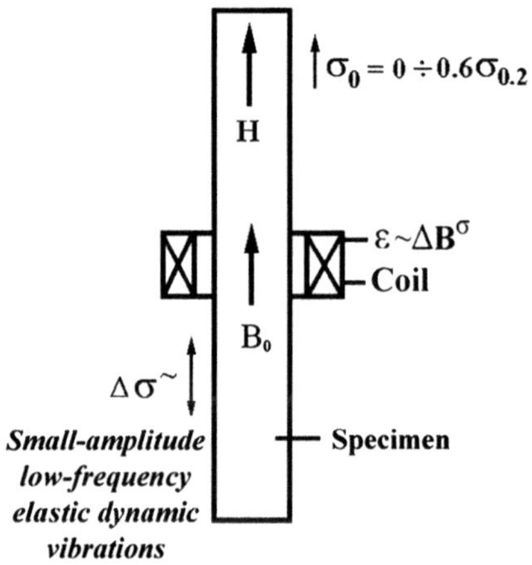

Fig. 4 Experimental scheme

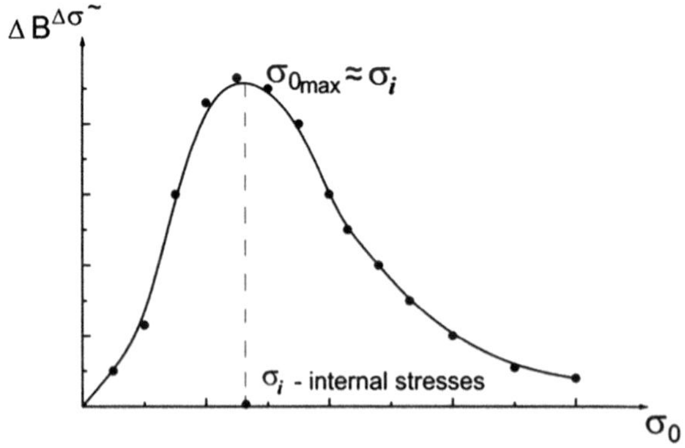

Fig. 5 The behaviour of magnetic induction caused in the specimen by elastic vibrations, $\Delta\sigma^{\sim}$, as a function of applied external stresses

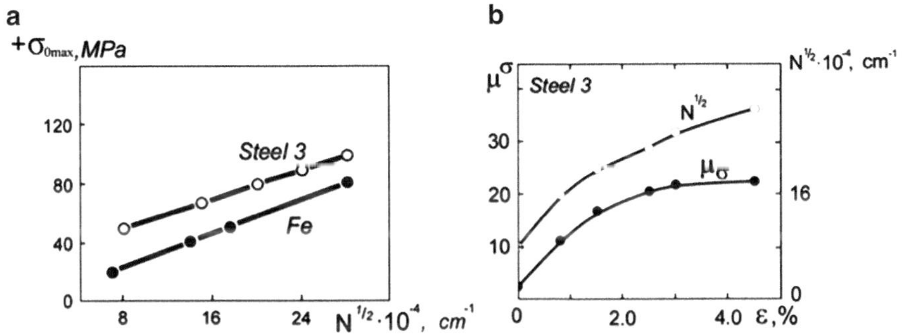

Fig. 6 A relation of σ_{0max} and magnetic permeability μ^σ caused by the displacement of 90° domain walls to dislocation density in iron and steel St3 (C ∼ 0.2%) suffering plastic deformation

It follows from [2, 3] that the load value at which the peak is reached correlates with the mean amplitude of internal stresses in differently deformed steels. Thus, according to [2, 3], the value of magnetic permeability μ^σ associated with the displacement of 90° DWs is determined as

$$\mu^\sigma \sim k\frac{\Delta B^{\Delta\sigma^\sim}}{\Delta\sigma^\sim B_0}. \qquad (1)$$

Experimental studies demonstrate (Fig. 6) that the values of σ_{0max} and μ^σ correlate well with dislocation density in iron and steel St3 under plastic deformation.

Another parameter enabling the microstress level in ferromagnetic materials to be estimated is magnetoelastic acoustic emission (MAE). According to current concepts, magnetoelastic acoustic emission is based on the magnetostrictive mechanism of excitation of signals produced by local sources of magnetostrictive strain at irreversible displacements of 90° DWs (Fig. 1b) and carrying information about changes in the magnetoelastic state of a ferromagnet.

When a ferromagnetic specimen is affected by a changing magnetic field, it suffers magnetization reversal along the magnetic hysteresis loop, and the irreversible displacement of 90° DWs causes elastic wave pulses (Fig. 1b), which are recorded by a piezotransducer as electric signals. After amplification, the number of jumps of electric signals of acoustic emission is recorded, or the rms values of emf U_{mae} of the envelope of all the jumps are measured. Materials with high values of magnetostriction show high acoustic activity in magnetization reversal.

As there is a relation between the stress state and magnetostriction, the parameters of magnetoelastic acoustic emission must also be related to the elastic properties of a ferromagnet, i.e., to the microstress level.

Figure 7 illustrates the behaviours of the rms value of U_{mae} and the microstress level as dependent on the tempering temperature for steel 34KhN3M (a) and on the amount of plastic strain for steel 40 (b). It follows from the analysis of Fig. 7 that U_{mae} correlates fairly well with the values of microstresses found by the X-ray

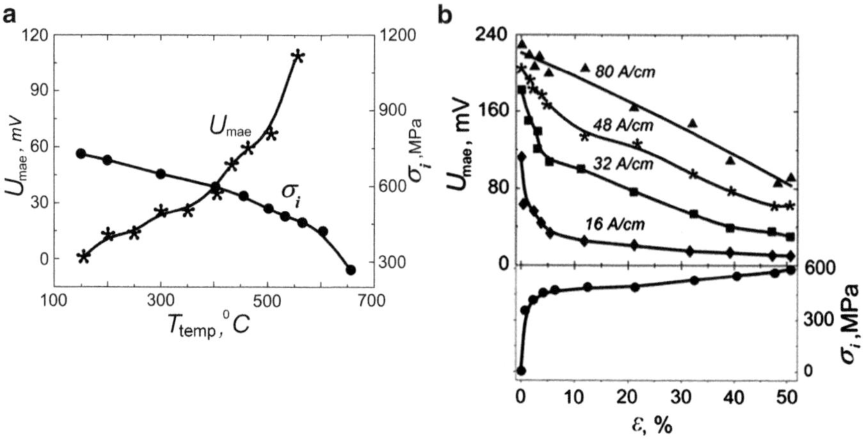

Fig. 7 The variation of the rms value U_{mae} and microstresses with tempering temperature for steel 34KhN3M (**a**) and with the amount of plastic strain for steel 40 (**b**)

technique. The MAE technique is employed to estimate the microstress level in thermally treated and elasto-plastically deformed ferromagnetic materials.

The third approach is magnetostrictive mutual transformation of elastic and electromagnetic waves in ferromagnets, i.e., electromagnetic-acoustic transduction (EMAT). EMAT signal parameters are also sensitive to the microstress level in a ferromagnet.

EMAT in steels was studied in a strong polarizing magnetic field by means of exciting passage coils and recording by the resonance technique to gain a standing elastic wave in the specimen. The EMAT signal amplitude was determined at elastic vibration resonance, as well as the resonance frequency enabling elastic wave propagation velocity to be calculated.

It is obvious from Fig. 8 that the EMAT signal amplitude and elastic wave propagation velocity decrease almost linearly with the growth of crystal lattice microdistortions caused by thermal treatment (quenching followed by tempering) of specimens made of carbon steel U8 (C $\cong$ 0.80%; Cr $\cong$ 0.23%; Mn $\cong$ 0.22%), low-alloy steel ShKh15 (C $\cong$ 0.98%; Cr $\cong$ 1.38%; Mn $\cong$ 0.32%), high-alloy steel R6M5 (C $\cong$ 0.84%; Cr $\cong$ 3.30%; W $\cong$ 5.92%; V $\cong$ 1.80%; Mo $\cong$ 4.83%), and this is suggestive of the applicability of EMAT parameters for estimating the level of microstresses.

4 Restoration of strain diagrams

The determination of the uniaxial stress–strain state of metals by magnetic techniques enables one, using different models of the mechanics of solids, to approach the problem of predicting the behaviour of the mechanical properties of a metal

Fig. 8 A relation between the EMAT signal parameters and crystal lattice microdistortions for thermally treated steels

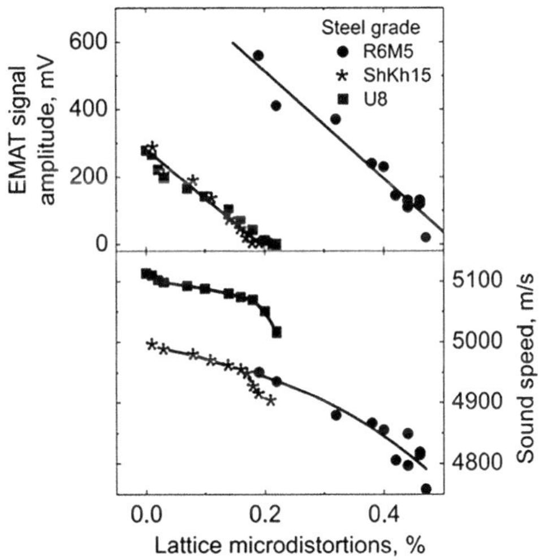

under deformation and estimating its residual life. It is proposed that magnetic methods of nondestructive testing [4] should be used for monitoring permanent strains in structural members. A technique has been developed for restoring strain diagrams by known relationships of standard mechanical properties of steel to measured magnetic characteristics.

The hardening curve is approximated by the equation

$$\sigma = \sigma_Y + A \cdot \Lambda^n, \tag{2}$$

where σ and σ_Y are current working stress and yield stress, $A = (\sigma_U - \sigma_Y)/\Lambda_{sh}^n$, $n = \sigma_U \cdot \Lambda_{sh}/(\sigma_U - \sigma_Y)$, $\Lambda_{sh} = 2 \cdot \ln(1+\delta)/\sqrt{3}$ is shear strain at the end of uniform tension.

If a correlation between standard mechanical and magnetic characteristics like $Y = a + bX$ is known, after substituting the values obtained from the correlations in (2), we arrive at a new deformation curve equation taking into consideration measured magnetic characteristics, (3). Thus, by the data obtained in [4], after substituting the correlation expressions $\sigma_Y \approx H_c$, $\sigma_U \approx H_c$ into (2), we have

$$\sigma = \sigma_Y(H_c) + A(H_c) \cdot \Lambda^{n(H_c)}. \tag{3}$$

The results are presented in Fig. 9.

The mechanisms of the deformation behaviour of the magnetic characteristics of low- and medium-carbon steels have been experimentally studied on all the portions of the stress–strain diagram up to necking under plane stress conditions in elastic-plastic tensile deformation.

In order to analyse the behaviour of the coercive force in tension, the "H_c versus amount of strain" curves were superposed on the corresponding stress–strain

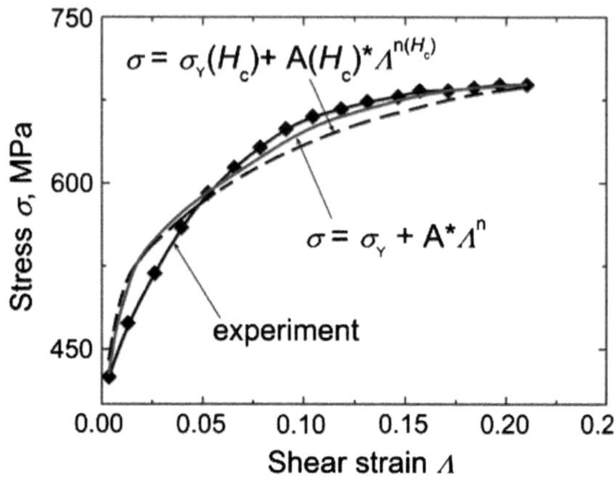

Fig. 9 Experimentally and analytically obtained tension diagrams, (3)

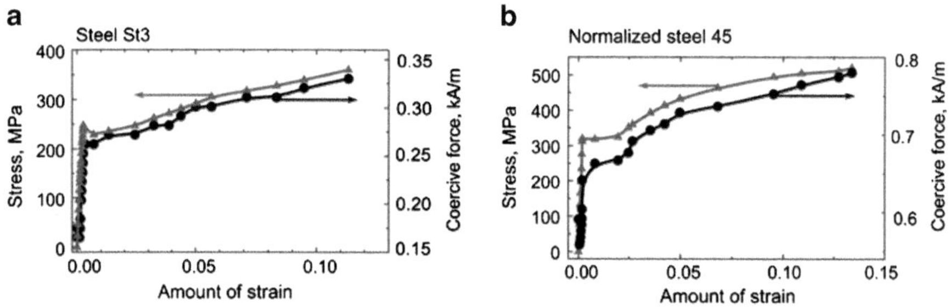

Fig. 10 Tensile stress and coercive force as dependent on the amount of strain for carbon steels

diagrams (Fig. 10). Three characteristic portions are obvious in the behaviour of the coercive force, namely, (1) the region of elastic strain; (2) the yield plateau and/or drop (sharp yield point); (3) the region of developed plastic strain.

The effect of stresses reaching and exceeding $\sigma_{0.2}$ leads to a collapse of the magnetic texture of stresses in the elastic strain region, and the coercive force in the plastic strain region is mainly affected by an increase in the density of dislocations and dislocation clusters ($H_c \sim N^{1/2}$), where N is dislocation density [5]) and the formation of a crystallographic strain texture.

As the stress $\sigma_{0.2}$ is approached, the coercive force of all the specimens tested grows significantly, the most active growth of H_c being observed prior to $\sigma_{0.2}$, rather than at $\sigma_{0.2}$, and this agrees with the results obtained in [6]. In steel St3 the coercive force stops growing as soon as the upper yield stress is reached, H_c remains constant or slightly decreases, and then, after passing the lower yield stress, it resumes growing, though less intensively (see Fig. 10a).

The investigation has offered a technique to restore stress–strain diagrams for homogenous steel products by coercive force, and this is a basis for magnetic monitoring of the stress–strain state of loaded steel members.

References

1. Gorkunov, E.S., Dragoshansky, Yu.N.: The Barkhausen effect and its utilization in the structuroscopy of ferromagnetic materials. I. The role of local defects and crystallographic orientation of ferromagnetics. Defectoskopiya. Rus. J. Nondestr. Test. (6), 3–23 (1999)
2. Nichipuruk, A.P., Gorkunov, E.S., a.u.: The effect of the dislocation structure produced by plastic deformation on the magnetic and magneto-elastic properties of iron and low-carbon steel. Phys. Met. Metallography, **74**(6), 593–597 (1992)
3. Nichipuruk, A.P., Gorkunov, E.S, a.u.: Magnetic and magnetoelastic properties of steel 30KhGSA after combination cold working and heat treatment. Rus. J. Nondestr. Test. **34**(4), 273–279 (1998)
4. Gorkunov, E.S., Bukhvalov, A.B., Kaganovich, A.Z., Rodionova, S.S., Durnitsky, V.N.: Studying the relation of mechanical and physical characteristics of continuously cast hot-rolled steel 45 to its structural parameters. Defektoskopiya (Rus. J. Nondestr. Test.) (6), 60–69 (1996)
5. Mikheev, M.N., Gorkunov, E.S.: Magnetic Methods of Structural Analysis and Nondestructive Testing, p. 252. Nauka, Moscow (1993)
6. Makar, J.M., Tanner, B.K.: The in situ measurement of the effect of plastic deformation on the magnetic properties of steel. Part I. Hysteresis loops and magnetostriction. J. Magn. Magn. Mater. **184**, 193–208 (1998)

On the Potential of Constraints in Nonlinear System Identification

Christoph Hametner and Stefan Jakubek

Abstract Local model networks (LMNs) offer a versatile structure for the identification of nonlinear static and dynamic systems. In this contribution the integration of prior knowledge in LMNs using equality constraints is presented. For that purpose, an equality constrained Generalised Total Least Squares (GTLS) algorithm for the local model parameter estimation of the LMN is presented. On the one hand GTLS parameter estimation helps to avoid bias in the estimates when some or all observations are subject to noise and on the other hand equality constraints allow to mathematically enforce desired system properties.

1 Introduction

The nonlinear system identification task is to build a model of a process from measured input and output data. When prior knowledge about the process is available, grey-box approaches allow to reduce the model complexity using physical (or other) insight into the nature of the system, [1]. In this context Local Model Networks (LMNs) are an established concept for a wide field of applications. The architecture of local model networks represents an excellent approach for the integration of various knowledge sources since they possess a good interpretability, [2].

In this contribution the integration of prior knowledge in local model networks using equality constrained Generalised Total Least Squares (GTLS) is presented. On the one hand implicit (qualitative) knowledge is taken into account which

C. Hametner (✉) · S. Jakubek
Christian Doppler Laboratory for Model Based Calibration Methodologies, Institute of Mechanics and Mechatronics, TU Vienna, Wiedner Hauptstr. 8-10/E325, A 1040 Vienna, Austria
e-mail: christoph.hametner@tuwien.ac.at; stefan.jakubek@tuwien.ac.at

H. Irschik et al. (eds.), *Advanced Dynamics and Model-Based Control of Structures and Machines*, DOI 10.1007/978-3-7091-0797-3_11,

determines fundamental dependencies of the input-output behaviour of the process, e.g. structure selection, main influence variables on the nonlinearity of the process or dynamic system behaviour. On the other hand quantitative knowledge which describes explicit dependencies between inputs and outputs is taken into account in order to improve the quality of the resulting model.

Local model networks interpolate between local models, which can easily be described using linear regression models. The approximation of nonlinear systems is achieved by partitioning of those regions of the input space where the nonlinearity is more complex, into smaller subdomains. The aggregation of such subdomains then yields a versatile description of the overall system.

The goal of the learning algorithm is to partition available training data into several subsets and to estimate the corresponding regression models, [3]. The model identification task can be divided into two subtasks: The *partitioning process* and the *local parameter estimation*, see Fig. 1.

The double arrows in Fig. 1 indicate that there is a strong bilateral interrelation between the partitioning process and the local model parameter estimation: The partitioning process is based on the local approximation quality (which inherently depends on the local parameter estimation) which in turn depends on the proper partitioning of subdomains.

In this work prior knowledge represented as equality constraints will be divided into *singular* and *functional* constraints:

- *Singular constraints:* These will typically arise from e.g. steady-state process measurements. For example when dynamic data of combustion engines are recorded on test benches there are usually steady-state operating data available, too. The adherence of dynamic models to these static input-output relations can be enforced by constraints.
- *Functional constraints:* These are known fundamental physical properties of the process to be identified. Examples would be e.g. differentiating behaviour, a static characteristic in a dynamic model, or a characteristic which must pass through the origin or must have an extremum.

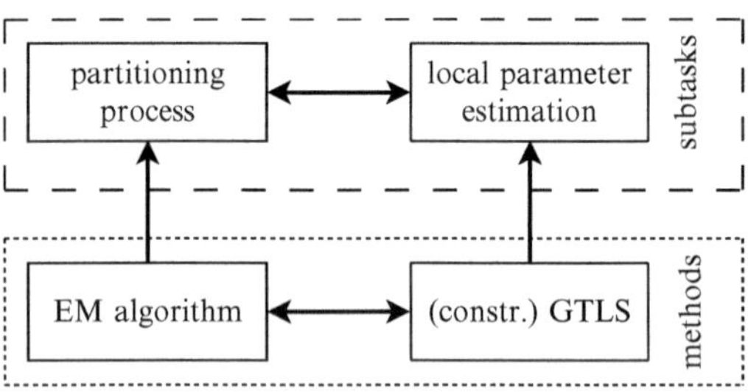

Fig. 1 Model identification subtasks and proposed methods

2 Architecture and construction of a local model network

Each local model (indexed by subscript i) of a local model network (LMN) consists of two parts: The local model parameter vector $\boldsymbol{\theta}_i$ and the validity function $\Phi_i(\tilde{\boldsymbol{x}}(k))$ which defines the region of validity in the partition space. The *local* estimate for the output $y(k)$ is obtained by

$$\hat{y}_i(k) = \boldsymbol{x}^T(k)\boldsymbol{\theta}_i, \tag{1}$$

where $\boldsymbol{x}^T(k)$ denotes the regressor vector. The *global* model output $\hat{y}(k)$ is obtained by a weighted aggregation of the local estimations $\hat{y}_i(k)$:

$$\hat{y}(k) = \sum_{i=1}^{L} \Phi_i(\tilde{\boldsymbol{x}}(k))\hat{y}_i(k), \tag{2}$$

where L denotes the number of local models.

The partition space is spanned by the so-called rule premises $\tilde{\boldsymbol{x}}$ with $\dim(\tilde{\boldsymbol{x}}) = p$. One of the main strengths of this approach lies in the fact that the input vector for the rule premises $\tilde{\boldsymbol{x}}(k)$ and the regressors $\boldsymbol{x}(k)$ can be chosen differently, [2]. Typically, the elements of the partition space $\tilde{\boldsymbol{x}}$ are chosen according to the expected nonlinear behaviour of the system based on prior system knowledge and experience. The main challenge in local model network parametrisation is to determine Φ_i in a way that the local estimate (1) sufficiently accurate describes the true process within the region of validity.

An exemplary model tree is depicted in Fig. 2, where each node corresponds to a split of the partition space into two parts and the free ends of the branches represent the actual local models with their validity functions Φ_i and the parameter vectors $\boldsymbol{\theta}_i$, cf. [4].

During training, in each iteration step the worst local model is replaced by two new local models which are then appended to the model tree. Thereby, an axis-oblique partitioning is performed by multi-objective optimisation using an Expectation-Maximisation (EM) algorithm, [5].

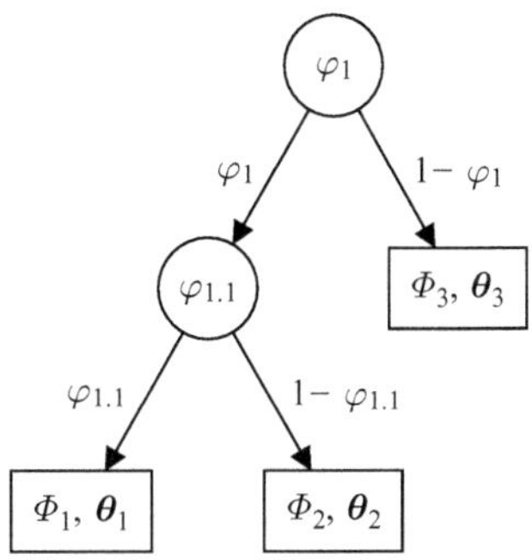

Fig. 2 Hierarchical discriminant tree

The goal of the iterative EM algorithm is to classify each data point in a way as to minimise the probability of missclassification. Usually, the prediction error is used for the partitioning process, [5]. When some or all input channels are subject to noise, a more general formulation of the residual is required. Accordingly, a residual for the GTLS estimator is defined for a proper statistical assessment of the residual error. This GTLS residual is used to discriminate between unsystematic errors from measurement noise and systematic errors from truncation.

3 Equality constrained GTLS parameter estimation

In many applications the inputs and outputs of a system are taken from measurements and are thus subject to noise. In this situation conventional parameter estimation methods suffer from the drawback that the parameter estimates are biased. Methods that are designed to cope with the case of noisy input and output channels are generally called "errors-in-variables methods", [6].

A proper parameter estimation in such cases can be obtained e.g. from Generalized Total Least Squares methods. Generally, GTLS algorithms are methods for *linear* parameter estimation, when some or all inputs are subject to noise [7]. The main advantage is that the GTLS algorithm as a generalisation covers both Least Squares (LS) and Total Least Squares (TLS) as special cases. For the integration in a local model network, a weighted GTLS (WGTLS) parameter-estimation algorithm that allows for individual weighting of data records is considered in this contribution.

3.1 *Weighted parameter estimation by GTLS*

Let $\boldsymbol{\Xi} \in I\!R^{N \times M}$ be a regressor matrix where every row contains all elements of $\boldsymbol{x}^T(k)$ except the constant offset and let $\boldsymbol{y} \in I\!R^{N \times 1}$ be the observation vector. For the reconstructions $\hat{\boldsymbol{y}}$, $\hat{\boldsymbol{\Xi}}$ a local affine model structure is chosen. For that purpose an augmented regressor matrix is defined:

$$\boldsymbol{W} = [\ \boldsymbol{y} \mid \boldsymbol{\Xi}\ ] \tag{3}$$

For GTLS parameter estimation $\boldsymbol{W}$ is recomposed by separating noisy components (indicated by subscript "n") and noise-free components (subscript "o"):

$$\boldsymbol{W} = \left[\boldsymbol{W}_n\ \boldsymbol{W}_o\right]. \tag{4}$$

The number of noisy and noise-free components is denoted by p_n and p_o, respectively. In the sequel, $\boldsymbol{W}$ (and its estimates) always refer to the definition given by (4).

GTLS algorithms constitute implicit estimators which are based on a low-rank approximation of the noisy observations.

Using a weighting matrix $\boldsymbol{Q}_i$ indexed by i with entries q_{ij} only along its main diagonal a weighted GTLS optimisation of parameters then yields

$$\|\boldsymbol{Q}^{1/2}(\boldsymbol{W}_n - \hat{\boldsymbol{W}}_n)\|_F^2 = \text{min!} \quad \text{subject to} \quad \hat{\boldsymbol{y}} \in \text{Image}(\hat{\boldsymbol{\Xi}}). \tag{5}$$

The solution is obtained by solving the following optimisation problem:

$$\boldsymbol{\beta}^T \underbrace{\begin{bmatrix} \boldsymbol{W}_n^T \boldsymbol{Q} \boldsymbol{W}_n & -\boldsymbol{W}_n^T \boldsymbol{q} & -\boldsymbol{W}_n^T \boldsymbol{Q} \boldsymbol{W}_o \\ -\boldsymbol{q}^T \boldsymbol{W}_n & s_q & \boldsymbol{q}^T \boldsymbol{W}_o \\ -\boldsymbol{W}_o^T \boldsymbol{Q} \boldsymbol{W}_n & \boldsymbol{W}_o^T \boldsymbol{q} & \boldsymbol{W}_o^T \boldsymbol{Q} \boldsymbol{W}_o \end{bmatrix}}_{\boldsymbol{\Omega}} \boldsymbol{\beta} = \text{min!} \tag{6}$$

subject to

$$\boldsymbol{\beta}^T \underbrace{\begin{bmatrix} \boldsymbol{I}_{p_n \times p_n} & \boldsymbol{0}_{p_n \times p_o+1} \\ \boldsymbol{0}_{p_o+1 \times p_n} & \boldsymbol{0}_{p_o+1 \times p_o+1} \end{bmatrix}}_{=\boldsymbol{\Phi}} \boldsymbol{\beta} = 1 \tag{7}$$

where $\boldsymbol{\beta}$ denotes an extended parameter vector.

The optimisation problem (6) and (7) can be solved using a Lagrange-function:

$$\mathcal{L}(\boldsymbol{\beta}, \lambda) = \boldsymbol{\beta}^T \boldsymbol{\Omega} \boldsymbol{\beta} - \lambda(\boldsymbol{\beta}^T \boldsymbol{\Phi} \boldsymbol{\beta} - 1) \tag{8}$$

which results in a generalised eigenvalue problem.

3.2 *Equality constraints*

Imposing additional equality constraints in the context of GTLS means that the parameter vector $\boldsymbol{\beta}$ has to fulfill additional conditions besides the requirements (6) and (7). Such constraints are introduced in form of additional regressors, termed $\boldsymbol{w}_c$. An equality constraint can then be expressed by the notion that $\boldsymbol{w}_c$ and its image are identical: $\hat{\boldsymbol{w}}_c(r) = \boldsymbol{w}_c(r)$. The index r indicates that there is usually more than one constraint to be imposed ($r = 1 \ldots n_c$). Since GTLS reconstructs only the noisy components the constraint can be formulated as follows: $\hat{\boldsymbol{w}}_{c,n}^T(r) = \boldsymbol{w}_{c,n}^T(r)$, where $\boldsymbol{w}_{c,n}$ and $\boldsymbol{w}_{c,o}$ are the partitions of $\boldsymbol{w}_c$ according to (4).

Using the extended parameter vector $\boldsymbol{\beta}$ a constraint is expressed as

$$\begin{bmatrix} \boldsymbol{w}_{c,n}^T(r) & -1 & -\boldsymbol{w}_{c,o}^T(r) \end{bmatrix} \boldsymbol{\beta} = 0. \tag{9}$$

A combination of all n_c constraints thus yields

$$\begin{bmatrix} \boldsymbol{w}_{c,n}^T(1) & -1 & -\boldsymbol{w}_{c,o}^T(1) \\ \boldsymbol{w}_{c,n}^T(2) & -1 & -\boldsymbol{w}_{c,o}^T(2) \\ \vdots & \vdots & \vdots \\ \boldsymbol{w}_{c,n}^T(n_c) & -1 & -\boldsymbol{w}_{c,o}^T(n_c) \end{bmatrix} \boldsymbol{\beta} = \boldsymbol{0} \tag{10}$$

or more compactly

$$\boldsymbol{S}_c \boldsymbol{\beta} = \boldsymbol{0}. \tag{11}$$

There are various approaches to solve the constrained parameter estimation problem outlined in this section, cf. [8, 9]. Exemplarily, the *null-space approach* is presented:

A suitable integration of (11) into the original optimisation problem (6) can be achieved if one ensures that $\boldsymbol{\beta}$ is parametrised such that (11) is always satisfied. This requires that $\boldsymbol{\beta}$ lies in the null-space of $\boldsymbol{S}$: $\boldsymbol{\beta} \in \mathcal{N}(\boldsymbol{S}_c)$. Let $\boldsymbol{N}_S$ denote a matrix with its column vectors spanning $\mathcal{N}(\boldsymbol{S}_c)$. Then a proper parametrisation of $\boldsymbol{\beta}$ is

$$\boldsymbol{\beta} = \boldsymbol{N}_S \boldsymbol{\mu} \tag{12}$$

with an arbitrary vector $\boldsymbol{\mu}$ and $\dim(\boldsymbol{\mu}) \leq p_n + p_o + 1$. Inserting (12) into the optimisation problem (6) and (7) leads to a generalised eigenvalue problem with reduced dimension $\dim(\boldsymbol{\mu})$:

$$\left(\boldsymbol{N}_S^T \boldsymbol{\Omega} \boldsymbol{N}_S - \lambda \boldsymbol{N}_S^T \boldsymbol{\Phi} \boldsymbol{N}_S\right) \boldsymbol{\mu} = \boldsymbol{0}. \tag{13}$$

4 Dynamic MISO example

In this section the identification of a nonlinear dynamic MISO system with three inputs is presented, Fig. 3. The structure is quite similar to the example in [10], where the system is used to validate another MISO identification scheme. The input u_1 and the output y are corrupted by measurement noise.

The differentiating behaviour of the transfer function $F_1(z^{-1})$ is assumed to be known a priori and will be imposed as a functional constraint in the training of the local model network. Additionally, a number of steady-state points were recorded which provide information on the static input-output relation of the system (singular constraints).

For the excitation APRB-signals were selected,with their amplitudes and bandwidths being designed such that the whole static and dynamic operating range of the MISO system can be covered. The input u_1 and the output y are corrupted by white noise. The signal-to-noise ratio is $\sigma_{ny}/\sigma_y = 0.002$ for the output and

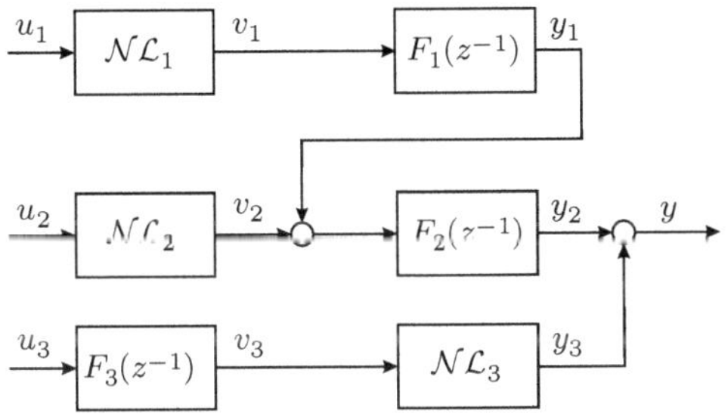

$$F_1(z^{-1}) = \frac{0.4066z^{-1}(1-z^{-1})}{1-1.605z^{-1}+0.6703z^{-2}}$$

$$F_2(z^{-1}) = \frac{0.2183z^{-1}+0.191z^{-2}}{1-1.605z^{-1}+0.6703z^{-2}}$$

$$F_3(z^{-1}) = \frac{0.1092z^{-1}+0.09552z^{-2}}{1-1.605z^{-1}+0.6703z^{-2}}$$

$$\mathcal{NL}_1 : v_1 = -\tfrac{1}{2}u_1 + \tfrac{1}{2}u_1^2 - u_1^3$$

$$\mathcal{NL}_2 : v_2 = -\tfrac{1}{2}u_2 - u_2^2 + \tfrac{1}{2}u_2^3$$

$$\mathcal{NL}_3 : y_3 = \tfrac{1}{2} - \tfrac{1}{2}v_3 - v_3^3$$

Fig. 3 Structure of the nonlinear MISO model

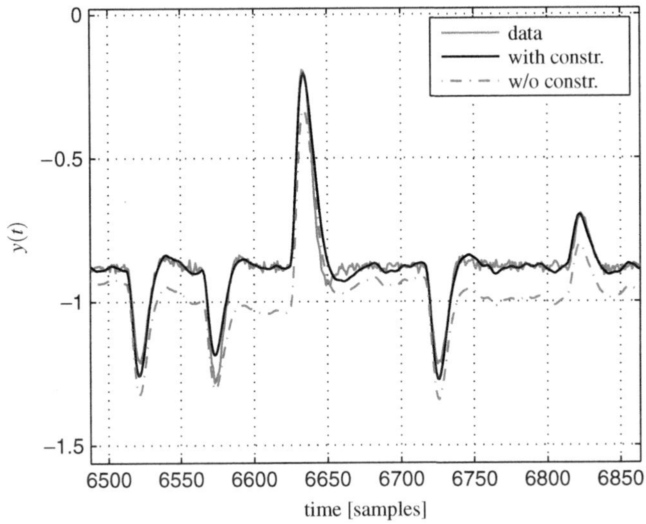

Fig. 4 Comparison of the model with and without equality constraints

$\sigma_{nu}/\sigma_{u_1} = 0.05$ for input u_1, respectively. The local model network training results in 8 local affine models, where the system order was chosen according to the known system, Fig. 3.

In Fig. 4 the simulation results of the model with no constraint information, the model with constraints and the actual process output are compared. In particular, u_2 and u_3 are chosen to be constant at the validation data record to verify the input-output relation of u_1 and y. It is clearly visible that the identification algorithm without constraints does not reproduce the differentiating behaviour of $F_1(z^{-1})$ exactly. Furthermore, a certain offset is monitored at the validation data record.

As expected, the differentiating behaviour is met exactly by model with integrated equality constraints and the offset vanishes due to the integration of steady-state constraints.

5 Conclusion

In this work the integration of prior knowledge by means of equality constrained GTLS parameter optimisation is presented. Prior knowledge of the nature of the system reduces the complexity of the model and parameter estimation itself. The GTLS algorithm yields consistent parameter estimates when some or all input channels are subject to noise and covers both LS and TLS as special cases. Together with a well tried local model network approach the constrained GTLS algorithm is applicable to nonlinear system identification tasks.

An equality constrained WGTLS algorithm is derived and allows for the integration of qualitative or quantitative knowledge about the process. Prior knowledge represented as equality constraints is divided into singular and functional constraints. The construction of the local models of the LMN is based on an Expectation-Maximisation algorithm using the GTLS residual.

The proposed concepts are validated by means of an illustrative example. The results show how equality constraints can be used to guarantee a priori known structural properties of the process in nonlinear dynamic system identification.

References

1. Ljung, L.: Perspectives on system identification. Proceedings of the 17th IFAC World Congress, Seoul, Korea, July 2008
2. Nelles, O.: Nonlinear System Identification, 1st edn. Springer Verlag, Berlin, Heidelberg, New York (2002)
3. Cherkassky, V., Ma, Y.: Multiple model regression estimation. Neural Networks, IEEE Trans. **16**(4), 785–798 (2005)
4. Breiman, L., Friedman, J., Olshen, R., Stone, C.: Classification and Regression Trees. Wadsworth and Brooks, Monterey, CA (1984)
5. Hametner, C., Jakubek, S.: Neuro-fuzzy modelling using a logistic discriminant tree. American Control Conference, 2007. ACC '07, pp. 864–869, July 2007
6. Soderstrom, T.: Errors-in-variables methods in system identification. Automatica, **43**(6), 939–958 (2007)
7. Van Huffel, S., Vandewalle, J.: Analysis and propoerties of the generalized total least squares problem $AX = B$ when some or all columns of A are subject to Error. SIAM J. Matrix Anal. Appl. **10**(3), 294–315 (1989)
8. Fletcher, R.: Practical Methods of Optimization, 2nd edn. Wiley, New York (1987)
9. Gill, P.E., Murray, W., Wright, M.H.: Practical Optimization. Academic Press, London (1981)
10. Treichl, T., Hofmann, S., Schroder, D.: Identification of nonlinear dynamic MISO systems with orthonormal base function models. In: Proceedings of the 2002 IEEE International Symposium on Industrial Electronics, vol. 1, pp. 337–342 (2002)

Vibration Control of Linear Elastic Beam Structures with Spatial Local Nonlinearities

Rudolf Heuer

Abstract Vibration problems of linear elastic structures with spatially localized nonlinearities are related to non-classically damped systems with single external devices. Such systems are characterized by the fact that their nonlinear behavior is largely restricted to a limited number of single points in the structure. The paper presents a semi-analytical procedure for analyzing the dynamic steady-state response of locally nonlinear beams under piezoelectric actuation, where special emphasis is laid on systems with single nonlinear spring or/and damping devices. The solution of the underlying boundary value problem is found by means of a problem-oriented decomposition in frequency domain, which has been adopted from methods for structures under multiple support excitations. Example problems are given for elastic beam structures with time-variant imposed piezoelectric curvature, where single supports are connected to external devices.

1 Introduction

Important elements of any structural control system are the transducers used for implementation of the control. Sensors and actuators provide the link between the controller and the mechanical system to be controlled and their design and implementation is of prime importance. At first, investigation was restricted to discrete vibration control, see, e.g., the survey on the technology of intelligent structures of Fuller et al. [1]. More recently, new material systems have made it feasible to carry out active control of distributed parameters for the sake of an efficient suppression of vibrations and structure-born sound. Studies in this and related systems have been made by Rao and Sunar [2] and others. Most frequently, the distributed

R. Heuer (✉)
Vienna University of Technology, Vienna A-1040, Karlsplatz 13/E2063, Austria
e-mail: rudolf.heuer@tuwien.ac.at

H. Irschik et al. (eds.), *Advanced Dynamics and Model-Based Control of Structures and Machines*, DOI 10.1007/978-3-7091-0797-3_12,

actuators in intelligent or smart type of structures are realized by embedding layers of piezoelectric materials, compare [3]. Such devices are often constructed in form of thin panels, plates or shells. Activation renders piezoelectrically induced strains that are imposed to generate the distributed input of the control system.

Classical Modal Analysis can be applied to linear systems if the corresponding damping matrix is proportional to both mass and stiffness matrices. Otherwise, e.g. in case of structures with single external damping devices, an alternative or approximate solution procedure for determining the dynamic response has to be chosen, compare Udwadia and Kumar [4]. Vibration problems of linear structures with spatially localized nonlinearities are related to those non-classically damped systems. The objective of this paper is to present an approximate procedure, i.e., a modified force-method, for analyzing the dynamic steady-state response of locally nonlinear beams under piezoelectric actuation. The solution is derived by means of a decomposition in frequency domain, which has been adopted from methods for structures under multiple support excitations. The main advantageous of this formulation is the fact, that by increasing the degrees of freedom considered for the structure, the numerical effort for solving the coupled system is extended only marginally since the nonlinearity stays "isolated".

2 Constitutive equations of a piezoelastic beam

In a Cartesian coordinate system, $x_i, i = 1,2,3$, using Einstein's convention of summation about repeated indices and assuming geometrically linearized conditions, the constitutive relations for a linear thermoelastic, anisotropic and piezoelectric solid can be formulated as the generalized Hooke's law,

$$\sigma_{ij} = c_{ijkl}^{E,\,\theta} \left(\varepsilon_{kl} - \varepsilon_{kl}^{*}\right); \qquad i,j,k,l = 1,2,3 \tag{1}$$

where σ_{ij}, ε_{kl} are stress and strain components. The fourth-order tensor of elastic constants, $c_{ijkl}^{E,\,\theta}$, (measured at constant strain and temperature θ) shows the following conditions of symmetry:

$$c_{ijkl}^{E,\,\theta} = c_{jikl}^{E,\,\theta} = c_{ijlk}^{E,\,\theta} = c_{klij}^{E,\,\theta}. \tag{2}$$

Due to the inverse piezoelectric effect, eigenstrains of the form

$$\varepsilon_{kl}^{*} = d_{mkl}^{\theta}\, \bar{E}_m, \tag{3}$$

are induced, where $\bar{E}_m$ denotes the components of the electric field vector. The third order tensor d_{mkl}^{θ} is symmetric with respect to k and l for $\varepsilon_{kl}^{*} = \varepsilon_{lk}^{*}$, thus

$$d_{mkl}^{\theta} = d_{mlk}^{\theta}. \tag{4}$$

For the present investigation, in which a large electric potential is applied to an isotropic beam structure, it is assumed that the electric field resulting from variation in stress and temperature (the so-called direct piezoelectric effect) is insignificant compared with the applied electric field. In that case the electric field vector is directed perpendicular to the (x_1, x_2)-plane, and in case of isotropic structures the components of the elastic tensor of (2) reduce to Young's modulus

$$c_{ijkl}^{E,\theta} = E, \tag{5}$$

and the in-plane eigenstrains, responsible for the flexural deformation, are

$$\varepsilon_{xx}^* = d_{zx}\,\bar{E}_z, \quad x \mathrel{\hat{=}} 1,\ z \mathrel{\hat{=}} 3. \tag{6}$$

Finally the stress distribution within the beam's cross section becomes

$$\sigma_{xx}(x, z_i;\, t) = E(z_i)\left[\varepsilon_{xx}(x, z_i;\, t) - \varepsilon_{xx}^*(x, z_i;\, t)\right]. \tag{7}$$

3 A modified force-method in frequency domain

Considering an undamped linear elastic single-span beam with a (nonlinear) device attached at the point $x = l_1$, see Fig. 1, the corresponding equation of motion reads

$$B\, w_{,xxxx}(x;t) + \mu\ddot{w}(x;t) = -B\,\kappa^*_{,xx}(x;t) - F(t)\,\delta(x - l_1), \tag{8}$$

where B and μ denote bending stiffness and mass per unit beam length, respectively. The imposed curvature,

$$\kappa^*(x;t) = \frac{1}{J_{yy}} \int_A z\,\varepsilon_{xx}^*\, dA, \tag{9}$$

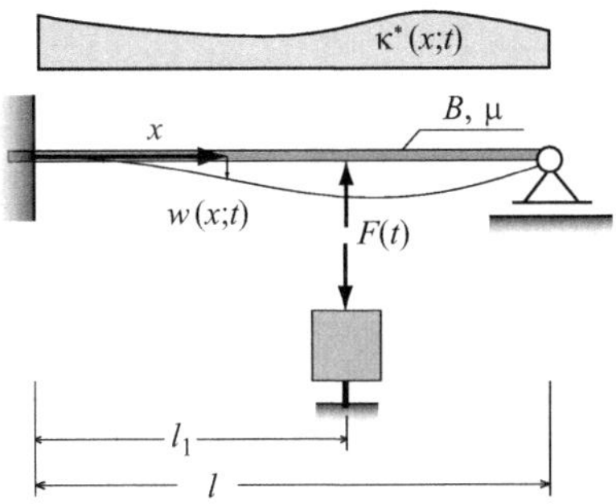

Fig. 1 Single-span beam with imposed curvature $\kappa^*(x;t)$ and external device at $x = l_1$

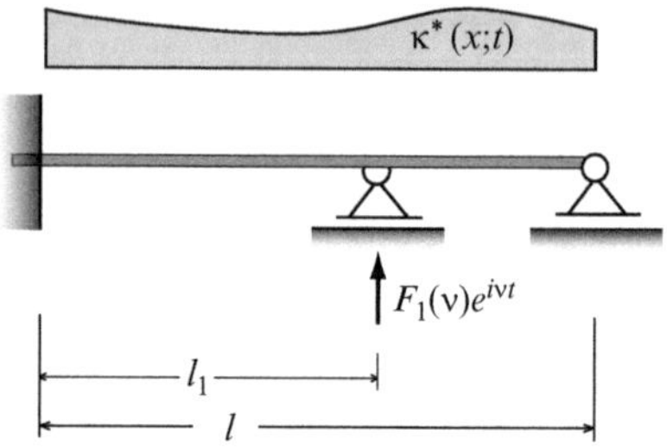

Fig. 2 Subsystem 1 of the underlying boundary value problem

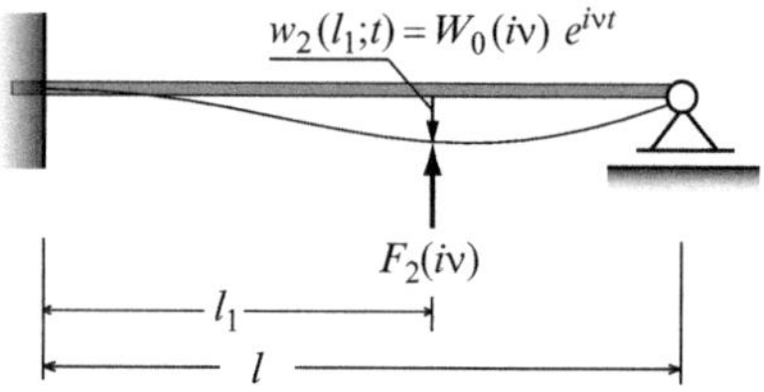

Fig. 3 Subsystem 2 of the underlying boundary value problem

is assumed to be separable in the space and time variables. J_{yy} stands for the axial moment of inertia of the cross sectional area A. Particularly, for steady-state excitation, the imposed curvature becomes

$$\kappa^*(x;t) = \psi(x)\,\kappa_0\,e^{i\nu t}. \tag{10}$$

In the first step of derivation, $F(t)$ is expressed as a sum of two separate forces,

$$F(t) = [F_1(\nu) + F_2(i\,\nu)]\;e^{i\nu t}, \tag{11}$$

where F_1 is the force generated due to $\kappa^*(x,t)$ at a support located at $x = l_1$ (subsystem 1), and F_2 arises due to an imposed deflection $w_2(x = l_1, t) = W_0(i\,\nu)\;e^{i\nu t}$ with no other external forces acting (subsystem 2), see Figs. 2 and 3.

In order to calculate the particular solution $W_p(x;\nu)$ associated to the differential equation

$$B\,W_{p,xxxx}(x;\nu) - \nu^2\mu W_p(x;\nu) = -B\,\psi_{,xx}(x)\,\kappa_0, \tag{12}$$

the total flexural response W_p is partitioned into a frequency independent (quasistatic) part, W_S, and into its complementary dynamic portion, W_D. Substituting that decomposition into (12), yields to the following set of two differential equations:

$$W_{S,xxxx} = -\psi_{,xx}\;\kappa_0, \tag{13}$$

$$B\,W_{D,xxxx} - \nu^2\mu\,W_D = \nu^2\mu\,W_S. \tag{14}$$

Equation (13) describes the static beam problem with imposed curvature distribution. It can be solved using classical methods, e.g. applying Maysel's formula, compare Parkus [5]. Its solution directly enters (14), which is solved by means of the influence function for the beam deflection, $\tilde{W}$:

$$W_D(x;\nu) = \nu^2 \int_l \tilde{W}(x,\xi;\nu)\,\mu\,W_S(\xi)\,d\xi. \tag{15}$$

The two parts of the dynamic force are evaluated from the following conditions:

$$W_p(l_1;\nu) - F_1(\nu)\,\tilde{W}(l_1,l_1;\nu) = 0, \tag{16}$$

$$F_2(i\,\nu) = -c_{dyn}(i\,\nu)\;W_0(i\,\nu), \tag{17}$$

where the complex value $W_0(i\,\nu)$ is yet unknown; c_{dyn} represents the (nonlinear) dynamic stiffness coefficient at $x = l_1$. Reformulation of (16) and (17) gives

$$F_1(\nu) = \frac{W_p(l_1;\nu)}{\tilde{W}(l_1,l_1;\nu)} = \frac{W_S(l_1) + \nu^2 \int_l \tilde{W}(l_1,\xi;\nu)\,\mu\,W_S(\xi)\,d\xi}{\tilde{W}(l_1,l_1;\nu)}, \tag{18}$$

$$F_2(\nu) = -\frac{W_0(i\,\nu)}{\tilde{W}(l_1,l_1;\nu)}. \tag{19}$$

The approximate solution for the response of the nonlinear system subjected to time harmonic excitation is obtained by means of equivalent linearization technique, where the external nonlinear device is consistently replaced by a set of linear elements, e.g., a generalized Kelvin-Voigt model with effective stiffness c_e and damping parameter r_e. Thus, in frequency domain, the following constraint yields the (nonlinear) equation for the unknown amplitude W_0:

$$F_1(\nu) + F_2(\nu) - (c_e + i\,\nu r_e)\,W_0(i\,\nu) = 0. \tag{20}$$

Finally, the total response in frequency domain can be cast in the form

$$W(x;i\,\nu) = W_p(x;\nu) + \frac{W_0(i\,\nu) - W_p(l_1,\nu)}{\tilde{W}(l_1,l_1;\nu)}\,\tilde{W}(x,l_1;\nu). \tag{21}$$

4 Applications

Two example problems of simply supported beams with single devices attached at $x = l_1$ will demonstrate the efficiency of the introduced formulation. For both applications it is assumed that the external loading is constantly distributed, $\psi(x) = 1$. Thus the quasistatic part becomes

$$W_S = W_S(x) = \frac{\kappa_0}{2} x^2 \left(\frac{l}{x} - 1 \right). \tag{22}$$

The complementary dynamic portion is calculated by means of the influence function for the deflection of a simply supported beam to give,

$$W_D = W_D(x; \nu) = \frac{\kappa_0 \nu^2}{l} \sum_1^\infty \frac{sin\, \alpha_n x}{\omega_n^2 \left[1 - (\nu/\omega_n)^2 \right]} \left[g_{n1} + g_{n2} \right], \tag{23}$$

where the abbreviations g_{n1} and g_{n2} stand for

$$g_{n1} = \frac{l}{\alpha_n^2} \left(sin\, \alpha_n l - l\, \alpha_n \cos \alpha_n l \right), \quad \alpha_n = \frac{n\pi}{l}, \tag{24}$$

$$g_{n2} = \frac{1}{\alpha_n^3} \left[2\left(1 - \cos \alpha_n l - l\, \alpha_n sin\, \alpha_n l \right) + l^2 \alpha_n^2 \cos \alpha_n l \right]. \tag{25}$$

The first part of the restoring force at $x = l_1$ becomes

$$F_1 = \kappa_0 \frac{\dfrac{l_1^2}{2} \left(\dfrac{l}{l_1} - 1 \right) + \dfrac{\nu^2}{l} \sum\limits_1^\infty \dfrac{(g_{n1} + g_{n2})\, sin\, \alpha_n l_1}{\omega_n^2 \left[1 - (\nu/\omega_n)^2 \right]}}{\dfrac{2}{\rho A l} \sum\limits_1^\infty \dfrac{(sin\, \alpha_n l_1)^2}{\omega_n^2 \left[1 - (\nu/\omega_n)^2 \right]}}, \tag{26}$$

and the second contribution, F_2, is determined from (19),

$$F_2 = -W_0 \left[\frac{2}{\rho A l} \sum_1^\infty \frac{(sin\, \alpha_n l_1)^2}{\omega_n^2 \left[1 - (\nu/\omega_n)^2 \right]} \right]^{-1}. \tag{27}$$

In the first example problem, a linear viscous damping device parameter r at $x = l_1$, is considered. In that case the equation for determination of the unknown complex amplitude becomes simply linear, compare (20),

$$F_1(\nu) + F_2(\nu) - i\, \nu r\, W_0(i\, \nu) = 0, \tag{28}$$

which can be solved analytically. Finally, the frequency response function (FRF) at any point of the beam axis follows from (21).

The second application deals with a nonlinear spring element, again attached to the simply supported beam at $x = l_1$, where the elastic restoring force is given in symmetric form by a cubic polynomial,

$$F_c = c\, W_0 \left(1 + \varepsilon W_0^2 \right). \tag{29}$$

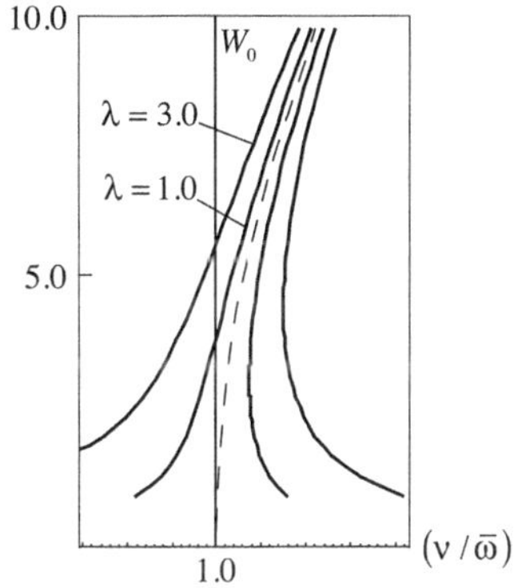

Fig. 4 Nonlinear FRF at $x = l/2$; $\lambda = \kappa_0 l$, $\varepsilon = 0.01$, $c = EJ\pi^4/l^3$, $\bar{\omega}^2 = 2(c + EJ\pi^4/2l^3)/\rho A l$

It is replaced consistently by

$$F_e = c_e\, W_0, \tag{30}$$

where the equivalent linearized system is found from Galerkin's procedure:

$$c_e = c\left(1 + \frac{3}{4}\,\varepsilon\, W_0^2\right). \tag{31}$$

In that case, the corresponding equation of the unknown amplitude is of cubic nature and it has to be solved numerically,

$$F_1 - \frac{3}{4}\,c\,\varepsilon\, W_0^3 - \left\{\left[\frac{2}{\rho A\, l}\sum_1^{\infty}\frac{(sin\,\alpha_n l_1)^2}{\omega_n^2\left[1 - (\nu/\omega_n)^2\right]}\right]^{-1} + c\right\} W_0 = 0. \tag{32}$$

The structure of (32) shows, that by increasing the degrees of freedom considered for the beam, the numerical effort for solving the coupled system is extended only marginally since the nonlinearity stays "isolated".

Figure 4, exemplarily, shows the result of the nonlinear FRF at the mid-point deflection for various load factors λ.

5 Conclusion

Piezoelectrically induced vibrations of elastic beams with local nonlinearities can be formulated by means of a modified force-method in frequency domain. The interaction force is expressed as a sum of two separate influences: the first develops due to the imposed piezoelectric curvature at the device's location, and the second contribution arises due to an imposed time-harmonic support excitation with no other external forces acting on the structure. It turns out, that by increasing the degrees of freedom considered for the structure, the numerical effort for solving the coupled system is extended only marginally since the nonlinearity stays "isolated".

References

1. Fuller, C.R., Elliott, S.J., Nelson, P.A.: Active Control of Vibration. Academic Press, London (1996)
2. Rao, S.S., Sunar, M.: Piezoelectricity and its use in disturbance sensing and control of flexible structures: a survey. Appl. Mech. Rev. **47**, 113–123 (1994)
3. Tzou, H.S., Gadre, M.: Theoretical analysis of multi-layered thin shell coupled with piezoelectric shell actuators for distributed vibration controls. J. Sound Vibration **132**, 433–450 (1989)
4. Udwadia, F.E., Kumar, R.: Iterative methods for non-classically damped dynamic systems. Earthquake Eng. Struct. Dyn. **23**, 137–152 (1994)
5. Parkus, H.: Thermoelasticity. Springer-Verlag, Wien (1974)

Energy Optimal Feedforward Control of a Cooling System

Anton Hofer and Filip Kitanoski

Abstract In hybrid electrical vehicles the cooling problem becomes more involved since additional components such as electrical drives, power electronics and batteries have to be cooled appropriately. Simultaneously the energy consumption used for the cooling task gets increased. Motivated by this fact an energy optimal control strategy for a cooling system which is forced by an electrically driven pump and an electrically driven fan is presented. It is shown how the control problem can be transformed to a mixed integer linear program which allows good approximations of the global optimal solution.

1 Introduction

The recent development in automotive industry is dominated by two goals – reduction of fuel consumption and CO_2 emissions of the vehicles. Hybrid electrical vehicles are regarded to gain essential progress in coping with these requirements especially for the next few years. But the realization of hybrid power train concepts also introduces some severe problems e.g. the mass of the vehicle increases by the additional components such as the electrical drive and the energy storage system and also the cooling system becomes more complex. Usually some electrically actuated cooling loops have to be implemented in order to provide the required cooling [7]. Furthermore sophisticated cooling management systems have to be developed such that the energy used for the cooling task is as low as possible. The latter fact is

A. Hofer (✉)
Institute of Automation and Control, Graz University of Technology,
Kopernikusgasse 24, 8010 Graz, Austria
e-mail: anton.hofer@tugraz.at

F. Kitanoski
Virtual Vehicle, Inffeldgasse 21a, 8010 Graz, Austria
e-mail: filip.kitanoski@v2c2.at

H. Irschik et al. (eds.), *Advanced Dynamics and Model-Based Control of Structures and Machines*, DOI 10.1007/978-3-7091-0797-3_13,

the motivation for the contribution presented in this paper. A profound rating of the performance of a cooling management system from the energetic point of view requires the knowledge of the minimum energy necessary for a certain cooling task. For this purpose a method for the energy optimal control of a cooling system will be developed. In order to achieve the global optimum solution the assumption that the driving conditions (e.g. a test cycle) are known *in advance* has to be posed.

2 Mathematical model of the cooling system and control problem

For the presentation of the proposed control method a simple cooling system for an electrical drive in a hybrid electrical vehicle is considered (see Fig. 1). It consists of an electrically driven pump, which forces the coolant mass flow $\dot{m}_W$ through the cooling loop. The cooling of the fluid is provided by a heat exchanger, where the air mass flow $\dot{m}_A$ can be influenced by an electrically actuated fan. The mathematical model for the thermal behavior of this system is derived by the so-called *lumped mass approach*. For this purpose the following five masses are introduced: M_M denotes the mass of the electrical drive, M_{W1} is the mass of the coolant in the pipe from the electrical drive to the heat exchanger, M_{W2} is the mass of the coolant from the heat exchanger to the electrical drive, M_S denotes the mass of the mechanical structure of the heat exchanger and M_A is the mass of the air in the heat exchanger. The temperatures of these masses are denoted by $T_M, T_{W1}, T_{W2}, T_S, T_A$. The thermal behavior of the system is mainly determined by the *adjustable* rotational speeds n_P and n_F of the pump and the fan respectively and by the heat rate $\dot{Q}_E$ produced by the operation of the electrical drive. Furthermore the cooling effect of the heat exchanger is also influenced by the environmental temperature T_E and the velocity of the vehicle v.

In order to simplify the mathematical description of this thermal system the following assumptions are made: The properties of the coolant fluid such as heat

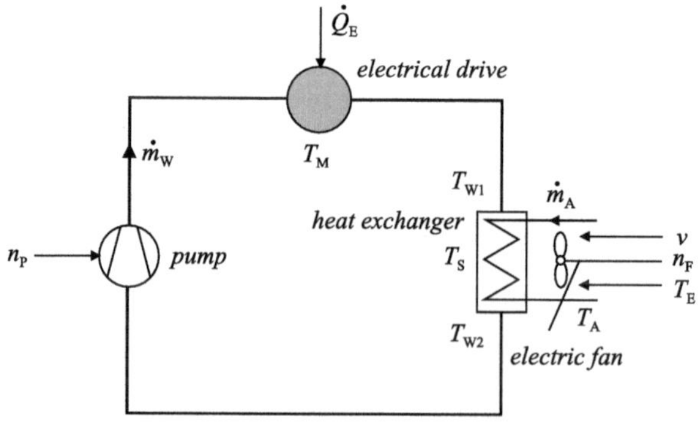

Fig. 1 Cooling system

capacity, density, viscosity are constant. The heat transfer between the pipes and the environment is neglected and only homogenous coolant and air mass flows in the system are considered. Then the following mathematical model can be derived

$$
\begin{aligned}
c_{p_M} M_M \frac{dT_M}{dt} &= -\alpha_{W1} A_{W1}(T_M - T_{W1}) + \dot{Q}_E \\
c_{p_W} M_{W1} \frac{dT_{W1}}{dt} &= \alpha_{W1} A_{W1}(T_M - T_{W1}) - c_{p_W}(T_{W1} - T_{W2})\dot{m}_W \\
c_{p_W} M_{W2} \frac{dT_{W2}}{dt} &= \alpha_{W2} A_{W2}(T_S - T_{W2}) + c_{p_W}(T_{W1} - T_{W2})\dot{m}_W \\
c_{p_S} M_S \frac{dT_S}{dt} &= -\alpha_{W2} A_{W2}(T_S - T_{W2}) + \alpha_A A_A(T_A - T_S) \\
c_{p_A} M_A \frac{dT_A}{dt} &= -\alpha_A A_A(T_A - T_S) - c_{p_A}(T_A - T_E)\dot{m}_A.
\end{aligned} \tag{1}
$$

The parameters $c_{p_M}, c_{p_W}, c_{p_S}, c_{p_A}$ are the heat capacities of the masses and α_{W1}, α_{W2}, α_A denote heat transfer coefficients and A_{W1}, A_{W2}, A_A are the corresponding contact areas. It is important that the heat transfer coefficients are depending on the mass flows which is taken into account in a simplified way by the following affine relations:

$$\alpha_{W1}(\dot{m}_W) = \alpha_{W1_0} + \alpha_{W1_1}\dot{m}_W \tag{2}$$

$$\alpha_{W2}(\dot{m}_W) = \alpha_{W2_0} + \alpha_{W2_1}\dot{m}_W \tag{3}$$

$$\alpha_A(\dot{m}_A) = \alpha_{A_0} + \alpha_{A_1}\dot{m}_A \tag{4}$$

Here $\alpha_{W1_0}, \alpha_{W1_1}, \alpha_{W2_0}, \alpha_{W2_1}, \alpha_{A_0}, \alpha_{A_1}$ are appropriately chosen constants. Furthermore it is assumed, that the rotational speeds n_P, n_F of the pump and the fan are not adjusted continuously by their electrical drives but are kept piecewise constant at some given levels. This assumption is realistic since it also reduces the costs for the realization of the cooling system. Thus neglecting the dynamics of the electrical drive of the pump the coolant mass flow $\dot{m}_W$ is also considered to be a piecewise constant time function taking only $m + 1$ distinct values a_i. It is represented by

$$
\begin{aligned}
\dot{m}_W(t) &= \sum_{i=0}^{m} a_i d_i(t) \\
d_i(t) &\in \{0, 1\} \\
\sum_{i=0}^{m} d_i(t) &= 1 \text{ for all } t.
\end{aligned} \tag{5}
$$

A somewhat different relation has to be introduced for the air mass flow $\dot{m}_A$, since it also depends on the velocity of the vehicle. It is assumed that the fan operates at $n + 1$ different stages which leads to

$$\dot{m}_A(t) = \sum_{i=0}^{n} (c_{0i} + c_{1i} v(t))\, \tilde{d}_i(t)$$

$$\tilde{d}_i(t) \in \{0, 1\}$$

$$\sum_{i=0}^{n} \tilde{d}_i(t) = 1 \text{ for all } t. \tag{6}$$

The constants a_i, c_{0i}, c_{1i} have to be determined from the pump and fan characteristics.

Based on this mathematical description we are ready to formulate the **control problem**: Given the initial values $T_M(0), T_{W1}(0), T_{W2}(0), T_S(0), T_A(0)$ and the time functions of the external inputs $\dot{Q}_E(t), T_E(t), v(t)$ for $0 \leq t \leq t_{end}$ e.g. by measurements from a test cycle, determine the binary valued functions $d_i(t)$ $i = 0, \ldots, m$ and $\tilde{d}_i(t)$ $i = 0, \ldots, n$ for $0 \leq t \leq t_{end}$ such that the electrical energy used for the operation of the pump and the fan for $0 \leq t \leq t_{end}$ is minimized and the constraints

$$T_M(t) \leq T_{\max} \tag{7}$$

$$|T_{W1}(t) - T_{W2}(t)| \leq \Delta T_{\max} \tag{8}$$

are satisfied for $0 \leq t \leq t_{end}$ with some prescribed bounds $T_{\max}$ and $\Delta T_{\max}$.

The first constraint poses a bound on the temperature of the electrical drive which must not be violated during the operation of the vehicle. The second constraint limits the difference of the temperature of the coolant at the inflow and the outflow of the heat exchanger in order to guarantee a proper operation of the heat exchanger. Considering the structure of the mathematical model (1–6) we are faced with an energy optimal feedforward control problem over a fixed time interval for a multivariable nonlinear timevariant system with input and state constraints and free terminal state. This control problem can be treated e.g. by the famous maximum principle of Pontryagin or by the dynamic programming method [8]. However since the prospect of success seems rather limited in both cases an alternative approach will be presented in the next section.

3 Transformation of the control problem into a mixed integer linear program

The proposed strategy to obtain a good approximation of the global optimum solution of this control problem consists of three steps: First a suitable time discretization is introduced in order to transform the control problem into a finite dimensional parameter optimization problem. Then using a well known method from linear programming the bilinear terms in the discrete time mathematical model are *equivalently* replaced by sets of linear inequalities with some binary variables. Finally the resulting mixed integer linear program (MILP) is solved by powerful algorithms [2].

Keeping in mind that the mixed integer linear program should become well tractable the time discretization should be carried out with a step size as large as possible. This goal is essentially limited by the fact, that (1) is a *stiff* system of differential equations because the thermal inertia of the air is much smaller than the thermal inertia of the other parts in the cooling system, i.e. $c_{PA}M_A \ll c_{PM}M_M, c_{PW}M_{W1}, c_{PW}M_{W2}, c_{PS}M_S$. In order to circumvent this problem the last differential equation in (1) is replaced by the algebraic equation

$$0 = -\alpha_A A_A(T_A - T_S) - c_{p_A}(T_A - T_E)\dot{m}_A. \tag{9}$$

Now introducing the definitions

$$\mathbf{x} := [T_M\ T_{W1}\ T_{W2}\ T_S]^T$$
$$\boldsymbol{\xi} := \left[\mathbf{x}^T\ T_A\right]^T \tag{10}$$

the model can be written as

$$\begin{bmatrix}\dot{\mathbf{x}}\\ 0\end{bmatrix} = \mathbf{A}_0\boldsymbol{\xi} + \mathbf{b}_0(t) + \sum_{i=1}^{m}\mathbf{A}_i\boldsymbol{\xi}d_i + \sum_{j=0}^{n}\left[\mathbf{b}_{1,j}(t) + \mathbf{B}_j(t)\boldsymbol{\xi}\right]\tilde{d}_j \tag{11}$$

with constant matrices $\mathbf{A}_0, \mathbf{A}_1, \ldots, \mathbf{A}_m$. The time dependence of the parameters $\mathbf{b}_0, \mathbf{b}_{1,0}, \ldots, \mathbf{b}_{1,n}, \mathbf{B}_0, \ldots, \mathbf{B}_n$ is determined by the known external input functions $\dot{Q}_E(t), T_E(t), v(t)$. Motivated by the fact that the control inputs are chosen as piecewise constant time functions with fixed values the time discretization is performed based on the state transition matrix as it is known from linear systems. With the notation $w_k := w(k\Delta t)$ for the value of a time function $w(t)$ at the time instant $k\Delta t$ the discrete time model can be derived as:

$$\begin{bmatrix}\mathbf{x}_{k+1}\\ 0\end{bmatrix} = \boldsymbol{\Phi}\boldsymbol{\xi}_k + \mathbf{h}_{0,k} + \sum_{i=1}^{m}(\mathbf{F}_i\boldsymbol{\xi}_k + \mathbf{h}_{i,k})\,d_{i,k} + \sum_{j=0}^{n}\left(\mathbf{G}_{j,k}\boldsymbol{\xi}_k + \mathbf{f}_{j,k}\right)\tilde{d}_{j,k} \tag{12}$$

A detailed presentation how the matrices $\boldsymbol{\Phi}, \mathbf{F}_i, \mathbf{G}_{j,k}$ and the column vectors $\mathbf{h}_{0,k}$, $\mathbf{h}_{i,k}$, $\mathbf{f}_{j,k}$ have to be calculated from the parameters of (11) is given in [5]. The validity of the proposed time discretization scheme has been proved by numerous comparisons of simulations carried out with the continuous time model (1–6) and the discrete time model (12). The step size Δt can be increased as much as the given time functions $\dot{Q}_E(t), T_E(t), v(t)$ are adequately represented by their corresponding sequences $\dot{Q}_{E,k}, T_{E,k}, v_k$.

The next step aims at the bilinear terms $\mathbf{F}_i\boldsymbol{\xi}_k d_{i,k}$ and $\mathbf{G}_{j,k}\boldsymbol{\xi}_k\tilde{d}_{j,k}$. Using a well known procedure [9] the bilinear terms can be *equivalently* replaced by a set of linear inequalities. Based on the realistic assumption that the set $\mathbb{X}$ of possible $\boldsymbol{\xi}_k$ is closed and bounded, upper and lower bounds for the products $\mathbf{F}_i\boldsymbol{\xi}_k$ can be determined

$$\mathbf{m}_{u,i} := \max_{\boldsymbol{\xi}_k \in \mathbb{X}} \mathbf{F}_i\boldsymbol{\xi}_k \quad \mathbf{m}_{l,i} := \min_{\boldsymbol{\xi}_k \in \mathbb{X}} \mathbf{F}_i\boldsymbol{\xi}_k \quad \text{for } i = 1,\dots,m \text{ and all } k \geq 0.$$

In these equations the notation means that the maximum and minimum operator is applied to each component of the vector $\mathbf{F}_i\boldsymbol{\xi}_k$. Next new vector valued variables $\mathbf{z}_{i,k}$ can be introduced which must satisfy the following linear inequalities[1] for $i = 0,\dots,m$ and all $k \geq 0$:

$$\mathbf{m}_{l,i} d_{i,k} \leq \mathbf{z}_{i,k} \leq \mathbf{m}_{u,i} d_{i,k} \tag{13}$$

$$\mathbf{F}_i\boldsymbol{\xi}_k - \mathbf{m}_{u,i}(1 - d_{i,k}) \leq \mathbf{z}_{i,k} \leq \mathbf{F}_i\boldsymbol{\xi}_k - \mathbf{m}_{l,i}(1 - d_{i,k}) \tag{14}$$

Now the important result can be easily verified that

$$\mathbf{z}_{i,k} = \mathbf{0} \text{ if } d_{i,k} = 0$$
$$\text{and } \mathbf{z}_{i,k} = \mathbf{F}_i\boldsymbol{\xi}_k \text{ if } d_{i,k} = 1$$

which has the useful consequence that the bilinear terms $\mathbf{F}_i\boldsymbol{\xi}_k d_{i,k}$ in (12) can be replaced by $\mathbf{z}_{i,k}$ and the inequalities (13,14). With (time dependent) upper and lower bounds $\tilde{\mathbf{m}}^k_{u,j}$, $\tilde{\mathbf{m}}^k_{l,j}$ for the products $\mathbf{G}_{j,k}\boldsymbol{\xi}_k$ the vectors $\tilde{\mathbf{z}}_{j,k}$ are selected next such that

$$\tilde{\mathbf{m}}^k_{l,j} \tilde{d}_{i,k} \leq \tilde{\mathbf{z}}_{j,k} \leq \tilde{\mathbf{m}}^k_{u,j} \tilde{d}_{i,k} \tag{15}$$

$$\mathbf{G}_{j,k}\boldsymbol{\xi}_k - \tilde{\mathbf{m}}^k_{u,j}(1 - \tilde{d}_{i,k}) \leq \tilde{\mathbf{z}}_{j,k} \leq \mathbf{G}_{j,k}\boldsymbol{\xi}_k - \tilde{\mathbf{m}}^k_{l,j}(1 - \tilde{d}_{i,k}) \tag{16}$$

are satisfied for $j = 0,\dots,n$ and all $k \geq 0$. Using the same reasoning as above we finally get the discrete time model in the simplified form

$$\begin{bmatrix} \mathbf{x}_{k+1} \\ 0 \end{bmatrix} = \boldsymbol{\Phi}\boldsymbol{\xi}_k + \mathbf{h}_{0,k} + \sum_{i=1}^{m} \mathbf{z}_{i,k} + \sum_{i=1}^{m} \mathbf{h}_{i,k} d_{i,k} + \sum_{j=0}^{n} \tilde{\mathbf{z}}_{j,k} + \sum_{j=0}^{n} \mathbf{f}_{j,k}\tilde{d}_{j,k} \tag{17}$$

augmented with the inequalities (13–16). This model opens the way to formulate the energy optimal control problem as a mixed integer linear program. Assuming that the behavior of the system is considered over the interval $0 \leq k \leq N$ of the time index (i.e. we have $t_{end} = N\Delta t$) the total energy E used for the operation of the cooling system can be expressed as

[1]Here the convention is used that inequalities between vectors should be interpreted component-wise.

$$E = \Delta t \sum_{k=0}^{N-1} \left(\sum_{i=1}^{m} p_i d_{i,k} + \sum_{j=1}^{n} \tilde{p}_j \tilde{d}_{j,k} \right) \tag{18}$$

where the constants p_i, $\tilde{p}_j$ denote the power of the pump and the fan at the specified stages. Now the control problem can be stated as mixed integer linear program: Determine the values of the binary variables $d_{i,k}$ $i = 0, \dots, m$ $k = 0, \dots, N-1$ and $\tilde{d}_{j,k}$ $j = 0, \dots, n$ $k = 0, \dots, N-1$ such that the energy (18) is minimized subject to the mathematical model (13–17) and the constraints

$$x_{1,k} \le T_{\max} \qquad k = 1, \dots, N$$

$$-\Delta T_{\max} \le x_{2,k} - x_{3,k} \le \Delta T_{\max} \qquad k = 1, \dots, N$$

$$\sum_{i=0}^{m} d_{i,k} = 1 \quad \sum_{j=0}^{n} \tilde{d}_{j,k} = 1 \qquad k = 0, \dots, N-1$$

The last equality constraints imply that at each time step k the binary variables $d_{i,k}$ $i = 0, \dots, m$ and $\tilde{d}_{j,k}$ $j = 0, \dots, n$ form so-called special ordered sets of type 1 (SOS1) [1]. The property that exactly one element of the set has to be non-zero can be exploited for speed-up of the solution process.

4 Example

The efficiency of the proposed method shall be demonstrated by an example, where the external inputs shown in Fig. 2 are taken from simulations of a section of the New European Driving Cycle (NEDC). It is assumed that the pump can produce the mass flows $0, 0.2, 0.5$ kg/s at the power levels $0, 80, 150$ W. Also the fan can be operated at three power levels $0, 250, 550$ W. The step size is chosen as $\Delta t = 2$ s and the bounds are selected as $T_{\max} = 50°\text{C}$ and $\Delta T_{\max} = 10°\text{C}$. A summary of the data used for the model can be found in [6]. The resulting mixed integer linear program has 900 binary variables, 9,904 real variables and 32,404 constraints. It was solved by the CPLEX/Tomlab solver [3, 4] on a 4Core-32GB Linux system within approximately 24 h CPU time. It is interesting to note that if the pump and the fan *always* operate at their highest level, we have $\max T_M \approx 48°\text{C}$ and a total amount of energy $E = 420$ kJ. So we may argue that there is no significant potential for a reduction of the energy in the present case. The solution of the MILP however reveals, that the required cooling task can be achieved selecting the pump and fan stages as it is shown in Fig. 3. In this case the energy consumption gets remarkably reduced to $E = 148.8$ kJ.

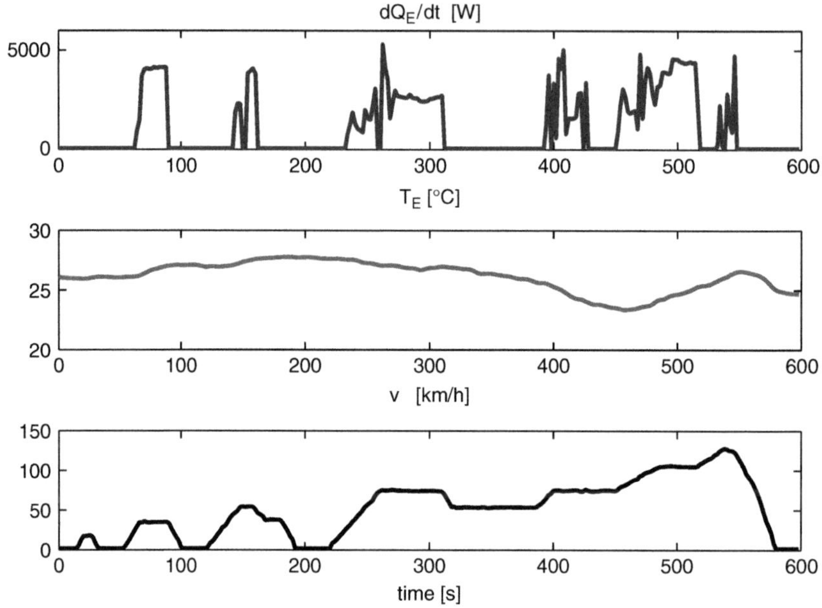

Fig. 2 External inputs from test cycle

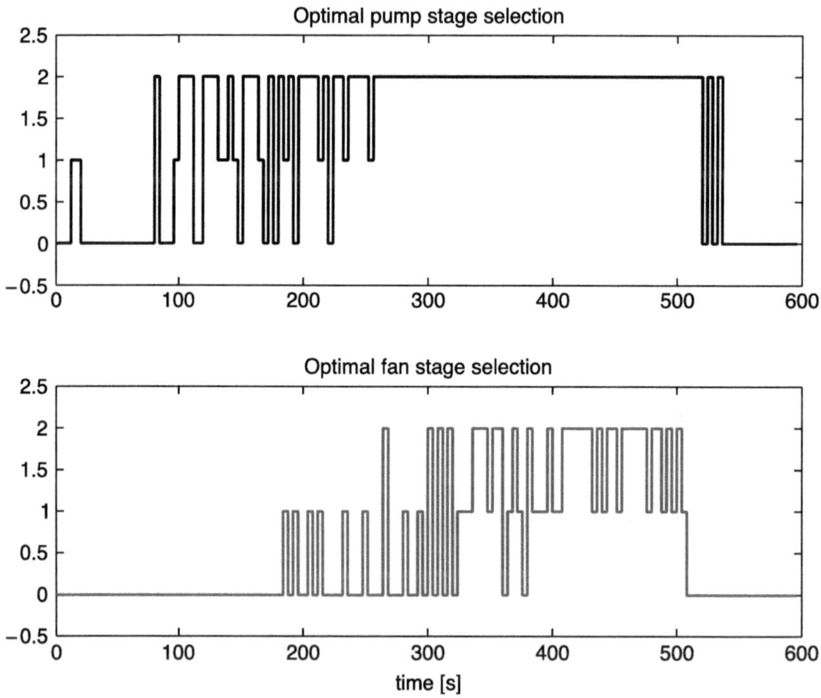

Fig. 3 Optimum selection of pump and fan stages

5 Conclusion

The transformation of the energy optimal control problem for a cooling system into a mixed integer linear program has been presented. This approach offers the possibility to obtain suitable approximations of the global optimal solution. The CPU time of 24 h required for the solution of the mixed integer linear program of the example above might seem unacceptable large. However in many practical cases the observation can be made, that near optimum solutions can be computed within one or 2 h but the proof of optimality can require days of CPU time. The proposed method is useful for the determination of cooling system layouts or for the development of cooling management systems. Of course a *feedback* solution of the energy optimal control problem would be preferable in the latter case. One possibility to determine such a controller might be obtained by identification applied to the optimum feedforward solution. This approach is the focus of actual investigations.

Acknowledgements This work was partly supported by the "COMET K2 Forschungsförderungs – Programm" of the Austrian Federal Ministry for Transport, Innovation and Technology (BMVIT), the Austrian Federal Ministry of Economics and Labour (BMWA), Österreichische Forschungsförderungsgesellschaft mbH (FFG), Das Land Steiermark, and Steirische Wirtschaftsförderung.

References

1. Beale, E.M.L., Forrest, J.J.H.: Global optimization using special ordered sets. Math. Program. **10**(1), 52–69 (1971)
2. Bixby, R.E., Rothberg, E.: Progress in computational mixed integer programming – A look back from the other side of the tipping point. Annals OR **149**(1), 37–41 (2007)
3. Holmstroem, K., Goeran, A.O., Edvall, M.M.: User's Guide for Tomlab 7. Tomlab Optimization Inc., San Diego, CA (2009)
4. IBM: User's Manual for CPLEX. IBM Corp. (2009)
5. Kitanoski, F.: Modellbildung, Simulation und optimale Steuerung des Kühlsystems eines Hybridfahrzeuges. PhD-thesis, Graz University of Technology, Austria (2010)
6. Kitanoski, F., Hofer, A.: Energy optimal control of a cooling system – limits of performance. Proceedings of the International Conference on Modelling, Identification and Control (MIC 2010), pp. 116–121 (2010)
7. O'Keefe, M., Bennion, K.: A comparison of hybrid electric vehicle power electronics cooling options. Proceedings of the IEEE Vehicle Power and Propulsion Conference, pp. 116–123. Arlington, TX, 9–12 September 2007
8. Papageorgiou, M.: Optimierung. R. Oldenbourg Verlag, München (1991)
9. Williams, H.P.: Model building in mathematical programming. Wiley, New York (1999)

Vibration Control of a Fluid in Micro- and Nanotubes

D.A. Indeitsev, A.K. Abramyan, and B.N. Semenov

Abstract In the present paper, we propose new equations describing the fluid behavior in nanochannels with its molecular structure. We consider the Poiseuille flow and the flow throw the channel with vibrating walls. The obtained results show that it is possible to describe the structural transformations in thin layers by using the continuum mechanics methods. We introduce new degrees of freedom of the material by introducing the second continuum, which plays the role of the arising new phase of state. In the models considered here, the properties of the new phase are determined by the influence of rigid boundaries with a different structure.

1 Introduction

Fluid flows in micro-and nanochannels are of great interest both from the standpoint of fundamental science and in practical applications [1, 2]. In this connection, the modeling of such flows is one of the most rapidly developing directions in hydrodynamics. The topicality of this modeling is also supported by the results of numerous experiments performed n the last two decades [1–4], which showed that there are significant distinctions of the fluid behavior predicted by classical continuum theories. It was found in these experiments that the fluid effective viscosity significantly increases in such volumes compared with its macroscopic value. Classical hydrodynamics, which does not take into account the fluid molecular structure, does not adequately describe the fluid flow in nanochannels of width equal or less that 50 molecular diameters. It is well known that the classical Poiseuille flow is described by the Navies-Stokes equations, and the velocity profile in this case is parabolic. Nevertheless, it was noticed long ago that some fluids begin to feel the

D.A. Indeitsev (✉) · A.K. Abramyan · B.N. Semenov
Institute for Problems in Mechanical Engineering, RAS, V.O. Bolshoy pr.61, 199178 St. Petersburg, Russia
e-mail: dmitry.indeitsev@mail2.ipme.ru

H. Irschik et al. (eds.), *Advanced Dynamics and Model-Based Control of Structures and Machines*, DOI 10.1007/978-3-7091-0797-3_14,

boundaries while flowing in sufficiently narrow gaps, which results in profile alterations. Moreover, numerous studies by Deryagin showed that on the boundary with a solid, fluids form layers with ordered structure determining the specific properties of the fluids in these layers [5]. In the present paper, to characterize such flows, we propose new equations describing the fluid behavior with its molecular structure.

2 Proposed model and main equations

The experimental data obtained in [1–4] permit formulating the main statement of the problem and constructing an appropriate mechanical model. In the opinion of the authors of the above-mentioned papers, the following structural changes arise in the interior of the nanosized layer: the forces of the structure interaction with the surfaces arranged the molecules of the medium filling the gap so that there appeared a solid-like ordered structure. This ordering depends not only on the intermolecular interaction of the medium particles but mainly on the molecule geometry and the surface of the contacting surfaces, which was conjectured in [2]. This influence of the surfaces on the molecule ordering led to the hypothesis that the solid like layer can have certain stiffness. As was noted in [3–5], one of the main effects of the boundary influence on the lubricant thin layer is the so-called polarization effect. The existence of such phenomenon is pointed out in [5]. If we rely on this fact in the description of the structure ordering and disordering phenomenon for fluids between two surfaces, then the basic equations of momentum and mass balance take the form

$$\left.\begin{aligned} G\frac{\partial^2 u_s}{\partial y^2} &= \rho_s\frac{\partial^2 u_s}{\partial t^2} + X_{sf}(u_s - v_f)\\ \mu\frac{\partial^2 v_f}{\partial y^2} &= \rho_f\frac{\partial u_f}{\partial t} + X_{fs}(v_f - u_s)\\ \frac{\partial \rho_f}{\partial t} + \nabla\cdot(\rho_f \mathbf{v}_f) &= X_{fs}\,,\ \frac{\partial \rho_s}{\partial t} + \nabla\cdot(\rho_s \dot{\mathbf{u}}_s) = X_{sf} \end{aligned}\right\} \tag{1}$$

where G and μ are the equivalent shear modulus of the structured medium and the equivalent viscosity of the "liquid" thus formed, $\dot{\mathbf{u}}_s$ and $\mathbf{v}_f$ are the velocities of the solid and fluid particles, ρ_s and ρ_f are the densities of the solid and fluid particles, X_{fs} is the rate of appearance (disappearance) of fluid particles at a reference point of the coordinate system.

We assume that, in a selected control volume of the medium, there is only one component of the combined medium at each time instant, which means that there are no direct interaction forces between the components. But, taken into account that there is an interface between the two media and this interface is variable, it is necessary to prescribe the interaction conditions on the boundary. The above equations describe the following scenario of events.

At rest, i.e., in the absence of any external load, the structured liquid is in the process of ordering as was described in [1]. This phenomenon is inhomogeneous with respect to the layer thickness, namely, the fluid particles in the central part of the layer are significantly less affected by the ordering forces as the particles on the boundaries with the surfaces. The three-dimensional MD (molecular dynamics) modeling confirms this.

2.1 *Poiseuille flow*

The above-proposed model of the fluid behavior was verified in the case of the Poiseuille flow. In this case, the system of (1) was solved. The following initial and boundary conditions were used:

$$\xi\big|_{t=0} = \xi_0(x, y), \quad p\big|_{x=0} = p_0, \quad p\big|_{x=L} = 0, \quad v_f\big|_{y=0} = v_f\big|_{y=H} = 0$$

where ξ is the fluid fraction in the elementary volume and $(1-\xi)$ is the solid fraction in the elementary volume. In this case, the densities ρ_f and ρ_s can be expressed in terms of their true values as $\rho_F = f_F/V = m_F V_F/(V \cdot V_F) = \rho_{F_0}\xi$ and $\rho_s = \rho_{s_0}(1-\xi)$, where V is the elementary volume and m and m are masses of a particle in each of the media.

To describe the fluid flow we use the two-component model. Namely we assume that the fluid in the channel is affected by the walls, i.e., has a possibility to be structured. The medium outside the channel is a usual molecular viscous fluid. The motion of the latter in the interior of the channel filled with a certain structured medium is similar to the flow through a "sieve" whose "feed through" cell dimensions significantly depend on the density of the ordered phase. We assume that, in the process of the fluid flow, the main resistance force is the reaction of fluid particle interaction with the structure cells, which is proportional to the difference of velocities of particles of the interacting components. At rest, without any applied external loads, the channel is filled with a medium which is ordered under the action of the channel walls. It is natural to assume that this phenomenon is inhomogeneous over the layer thickness, namely, the medium particles in the central part of the layer experience lesser influence of the walls than the particles on the boundary with the surfaces. In the present paper, we consider some specific cases in which the influence of the walls is such that, as a rule, the structures near the walls are more concentrated than those in the middle. The stressed state of the ordered medium is modeled as the pure shear stress. The so-called molecular liquid is fed into the channel, and this liquid interacts with the structure. This interaction force mainly depends on the density of the ordered structure, and hence on the so-called flow section through which the incoming particles can pass, and hence the force of the two media interaction is the largest. Thus, this interaction of two media, like the viscous friction force, depends on the difference of their particle velocities: it is the larger the higher the velocity of motion of one component relative to the other. In this case, it is assumed that this dependence is of a linear character. Obviously, as the incoming particles of the molecular liquid move with an input velocity greater than a certain value, the liquid has the tendency to pass freely through the immovable structured medium with possible separation of particles of the latter. This means that if the input pressure is sufficiently high, then the liquid medium can "destroy" the structure where it exists. Otherwise, if the pressure is insufficient, then the velocities of the applied particles are small, the structure density increases, and the flow rate of the constantly incoming liquid decreases. If the incoming liquid particles are sufficiently slow, then the structurization continues until complete sedimentation of liquid particles, i.e., the channel is "choked up" and the liquid cannot pass through it anymore. It is important to note that, in the equations of mass balance, there arise source terms determining the rate of transformation of liquid-like particles into solid-like particles and conversely. We assume that the sedimentation rate must be proportional to the particle concentration in the liquid, and the separation rate must be proportional to the structured medium concentration. Obviously, as the number of the liquid particles decreases, the number of solid-like particles increases, and hence the inverse process may occur. Because an ordered structure is formed for certain pressure and velocities, the flow rate through each cross-section of the channel decreases in the time. We emphasize that there exists such a regime at which there is a gradual separation of structured particles in the channel except for the boundary layers. It is the study of the above phenomena that resulted from the initial and boundary-value problem posed above. The main effect considered in this problem is the phenomenon of molecular liquid sedimentation on the structure, which may result in the so-called "choking" effect. Therefore, the final system of balance relation takes the form

$$\nabla \cdot \left\{ -\mathbf{I}p + \mu \left[\nabla \mathbf{v}_f + (\nabla \mathbf{v}_f)^T \right] \right\} - k_0 g(\xi)\, \mathbf{v}_f = \rho_f \frac{\partial \mathbf{v}_f}{\partial t}$$

$$\frac{\partial \xi}{\partial t} = J, \quad \nabla \cdot (\xi \mathbf{v}_f) = 0 \tag{2}$$

where J is the rate of sedimentation and separation of liquid particles at the checkpoint of the reference system has the form

$$J = \begin{cases} -k_1 \xi\,, \ \left|\mathbf{v}_f\right| < v^* \\ k_2(1-\xi)\,, \ \left|\mathbf{v}_f\right| > v^* \end{cases}, \quad g(\xi) = \frac{1}{D_0/(1-\xi) - D_1} \tag{3}$$

where D_0 is the characteristic open area dimension of the structured cell, D_1 is a coefficient of the Taylor expansion of $D_0(\rho_s)$, k_0 is the constant obtained from experimental data, $\mathbf{v}_f$ is the fluid velocity vector, k_1 and k_2 are constants obtained experimentally, and $\mathbf{v}^*$ is a certain critical velocity. The source term is assumed to have such a form because of the above-described scenario of events. The interaction forces between particles have the following form:

$$R_{msx} = \frac{k_s n_s}{D(n_s)} \mathbf{V}_m, \quad R_{msx} = -R_{smx}$$

where $D(n_s)$ is the cell characteristic open area dimension, and n_m and n_s are the densities of the structure and molecular liquid particles in the elementary volume. The total number of particles is equal to

$$N_s = \int_V n_s dV, \quad N_m = \int_V n_m dV$$

The conditions at the initial time moment are chosen as follows: they correspond to the assumption that the structure is more ordered near the walls and less ordered near the channel center. As follows from expression (3), the quantity $g(\xi)$ depends on the molecular liquid fraction in the entire volume ξ and on the values of the coefficients D_0 and D_1. As $\xi \to 1$, the quantity $g \to 0$. In this case, the quantity $k_0 g(\xi)\mathbf{v}_r$ contained in the first equation of the system (2) tends to zero, and the equation itself tends to the classical form. The solution of the above-posed problem for different sets of parameters confirms the qualitative applicability of the two-component model for describing the effect under study. The choking effect was investigated for different types of initial ordering of the medium, i.e. for different characters of the wall action and for two types of the source term. Now we consider the diagrams obtained using the mathematical models. The results given below show that the wall material structure itself significantly affects the liquid flow. In the first computer experiment, we consider a channel whose walls affected the incoming liquid so that the medium was structured mainly near the walls and significantly less near the center of the channel. Prescribing a certain initial pressure at which the liquid particles were incoming and the other necessary parameters, we observed a regime in which the velocity profiles and the concentration of each of the media had the form shown in Fig. 1. The character of the observed regime is shown in Fig. 1 for H=100 nm and L=200 nm.

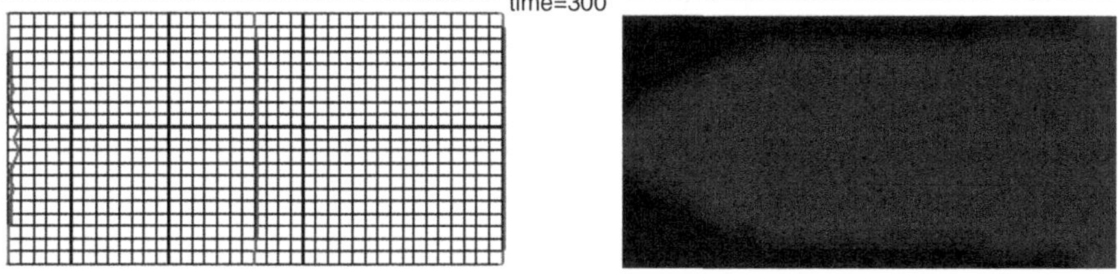

Fig. 1 The "chocking" of the channel

In the second computer experiment, we considered a channel whose walls affected the incoming liquid so that the medium was structured very strongly near the input and the walls and significantly less near the center of the channel. The diagrams for the two above – described experiments clearly illustrate the chocking regime and show that the input pressure was insufficient in these experiments, which is testified by the decrease in the flow rate of the incoming media. The computations showed that by increasing the excess pressure in the channel, one can obtain the converse effect, namely, the structured media becomes disordered. We did not calculate how the characteristic open area dimension affects the flow rate. In the third computer experiment, we considered a channel in which the structured medium distribution was similar to the preceding distribution but the source term in the mass balance equation had the form

$$J = -k_1(A - \rho_s)\rho_f H(v^* - v_F) + k_2\rho_s(B - \rho_f)H(v_F - v^*) \tag{4}$$

The first term describes the sedimentation (ordering) at a rate proportional to the quantity of the matter in the liquid and bounded by the quantity of the already ordered medium till the saturation A. The second term describes the converse transition. The results obtained confirm that there is a blocking effect, which is illustrated by an increase in solid-like phase concentration and a decrease ib liquid flow rate for a certain pressure regime. We note that, under the assumption of strong effect of the walls strongly on the medium in the channel, this model more clearly illustrates the action of these forces on the process of structurization. This phenomenon, under the assumption that the medium structurization must decrease from the walls towards the center of the channel, is most precisely described by using the source term in the form (4). The third experiment clearly shows that the particles begin to settle near the already structured medium.

2.2 Vibration influence on the fluid flow in the channel

In this part we consider the possibility of the vibration control of the fluid flow in the channel. Let us consider the oscillation of the channel walls with the small amplitude, and the influence of this vibration on the behavior of the fluid which consist both structured and liquid phases. The influence of the fluid viscosity is neglected in the following solution of the problem.

The governing equations in this case have the form

$$-\nabla p = \alpha(n_s)\mathbf{v}\,|\mathbf{v}| + \rho_f\dot{\mathbf{v}}_f, \quad 0 \le y \le h_0 + \tilde{h}(t), \;\; 0 \le x \le L$$

$$\alpha(n_s) = \frac{\kappa(n_s)}{D_0 - D_1 n_s}; \quad \kappa(n_s) = \kappa_0 + \kappa_1 n_s + ...$$

$$\mathbf{v}_f \cdot \mathbf{n}\big|_{y=h(t)} = \dot{h}, \quad \mathbf{v}_f \cdot \mathbf{n}\big|_{y=0} = 0$$

where κ_0 is coefficient which depends on the walls roughness. In general case $\kappa(n_s)$ depends on the density of structured particles of the liquid. The influence of the structured liquid movement will not be taken into consideration in the following solution of the system. Averaging the equation of the incompressibility over the channel height leads us to the following relation between the velocity of the fluid flow and vibration characteristics of the upper channel vibrating wall

$$v_f = \dot{h}(t)x + c(t), \quad h = h_0 + \tilde{h}\sin\omega t$$

Particle of incoming molecular liquid may increase the number of the structured liquid particles, and on the other hand when the velocity of the flow is higher than its critical magnitude then the number of the structured particles may decrease. The equation of particle s balance has the form

$$\frac{dn_s}{dt} = J(n_s, n_f), \quad \frac{dn_f}{dt} = -J(n_s, n_f) \tag{5}$$

The source term in this equation can be taking according (4).

Assume for simplicity that evolution process which described by the (5) is the slow process. Because of the slow velocity of the evolution process we assume that x has the constant value during the one period of oscillations. In this case one can get the following expression for the square of the average value of the flow velocity over one period

$$c^2(t) = \left[\Delta p - \frac{2}{3}\kappa(n_s)\left(\frac{\dot{h}}{h}\right)^2 L^3\right]\frac{h^2}{2\kappa(n_s)L}$$

If we take the condition that the magnitude of the flow rate over one period must be equal to zero than we comes to the expression of the critical value of the structured liquid particle density:

$$\hat{n}_s = \frac{3\Delta p h_0^2}{3\Delta p h_0^2 + \kappa_0 L^3\omega^2 h_0/D_1} \cdot \frac{D_0}{D_1}$$

Notes that the critical value of n which corresponds to the chocking of channel for the case of fixed walls is equal to $n_s^* = D_0/D_1$.

The vibration decreases this critical number, and essentially depends on the vibration frequency of the wall. Note, that this result needs some corrections which will take into account the influence of vibrations on the process of the particle exchange between the structured and molecular types of the liquids.

3 Conclusion

We propose a mathematical model of a fluid flow in a plane nanochannel, which is caused by the vibration of one of walls and Poiseuille flow. The obtained results show that it is possible to describe the structural transformation in thin layers by using the continuum mechanics methods. We introduce new degrees of freedom of the material by using the second continuum, which plays the role of the arising new phase of the state. In the models considered above, the properties of the new phase are determined by the influence of rigid walls with different structure. We also note that the two-component model can describe quite well such effects as the flow "chocking" and the "destroyed" layer reconstruction. This is because the source terms are introduced in the equation of the particle number balance of one or the other component. Depending on the scenario of the event in the material, it is quite possible to control and describe its state by using a suitable source term.

Acknowledgements The paper is written with the financial support of Russian Foundation of Basic Investigations, grants 08-01-00691-a and 10-01-00814-a.

References

1. Israelashvili, J.N., McGuiggan, P.M., Homola, A.M.: Dynamic properties of molecularly thin liquid films. Science **240**, 189–191 (1988)
2. Israelashvili, J.N.: Intermolecular and surface forces, 2nd edn. Academic Press, New York (1992)
3. Vinogradova, O.I.: Slippage of water over hydrophobic surfaces. Int. J. Miner Process. **56**, 31–40 (1999)
4. Thompson, P.A., Robbins, M.O.: Shear flow near solids: epitaxial order and flow boundary conditions. Phys. Rev. A. **41**, 6830 (1990)
5. Deryagin, B.V., et al.: Properties of fluids in thin quartz capillaries. Surface Forces in Thin Films and Colloid Stability, pp. 90–94. Nauka, Moscow (1974)

A Model Reduction Technique for High Speed Flexible Rotors

Hans Irschik, Manfred Nader, Michael Stangl, and Hans-Georg von Garssen

Abstract The present paper deals with a problem-oriented novel model reduction technique for elastic rotors with a high and/or rapidly changing axial speed. Numerical results are presented for various run-up simulations with different values of unbalance mass and internal damping ratio. An important result of this study is that the time-derivatives of the tilting angles and the flexible coordinates should not be considered as small when studying resonance and stability phenomena.

1 Introduction to a novel non-linear rotor model

The present contribution exemplarily reviews and extends some of the research, which has been performed in the last years by our group concerning the modelling and simulation of the motion and deformation of rotors, with a high and/or rapidly changing axial angular speed. This research has been undertaken at the Institut für Technische Mechanik of Johannes Kepler University in Linz, together with the Linz Center of Mechatronics and Corporate Technology of Siemens in Munich in the framework of the peer-reviewed Austrian Center of Competence in Mechatronics

H. Irschik (✉)
Institute for Technical Mechanics, Johannes Kepler University Linz,
Altenbergerstr. 69, 4040 Linz, Austria
e-mail: hans.irschik@jku.at

M. Nader · M. Stangl
Linz Center of Mechatronics GmbH, Altenbergerstr. 69, 4040 Linz, Austria
e-mail: manfred.nader@lcm.at; michael.stangl@lcm.at

H.-G. von Garssen
Siemens AG, CT PS 8, Otto-Hahn-Ring 6, 81730 Munich, Germany
e-mail: hans-georg.garssen@siemens.com

H. Irschik et al. (eds.), *Advanced Dynamics and Model-Based Control of Structures and Machines*, DOI 10.1007/978-3-7091-0797-3_15,

(ACCM). The main goal of this research has been to develop accurate and low order mathematical models, in order to provide a rational basis for the model based control of the mechanical behaviour of high speed rotors.

The following reasons for developing novel formulations in the more or less classical field of rotating bodies are mentioned. Classical rotor dynamics represents a highly valuable formulation given that the axial rotation speed is not too high, and the transverse displacements and velocities of the rotor axis are small. This results in linear equations of motion, see e.g. Krämer [1], Gasch et al. [2]. For higher axial rotation speeds, or if the axial rotation speed is rapidly changing, or for higher transverse velocities of the axis, as well as for the case that the rotor as a whole is mounted on a rapidly moving vehicle, a non-linear coupling between the axial rotation speed and the small or large rigid body motion of the vehicle, which carries the bearings of the rotor, is to be taken into account. Even when the displacements of the bearings are small, the transverse velocities of the axis may be large. Moreover, an accurate formulation is essential for considering the coupling between the rigid-body motion and the small elastic deformation of the rotor. This involves the necessity of considering the complex geometric shape and the material inhomogeneity of the rotor in an accurate manner. While the above mentioned non-classical modelling aspects in principle could be taken into account by using modern multi-body dynamics computer codes, such as ADAMS [3], the equations of motion would not be at disposal in an analytic form. However, the model based design of control algorithms for non-linear problems even today requires a sufficiently low number of equations of motion to be given in analytic form. That is why proper model reduction techniques are needed. Therefore, we have decided to derive an own code, which utilizes and extends the floating-frame-of-reference formulation (FFRF) of multi-body dynamics, and which is applied to the single-body problem of a rotating elastic body. For the classical FFRF, see the book of Shabana [4].

In our own formulation, we use a co-rotating rigid rotor as reference configuration, which allows introducing linear elastic modes of the non-rotating elastic rotor as Ritz approximations. The position vector of the origin of a Cartesian coordinate system attached to the co-rotating rigid rotor and three suitable Bryant angles are used as rigid-body coordinates, and free elastic modes of the non-rotating rotor are considered as Ritz approximations for the elastic deformation of the rotor, which thus is described by an additional set of flexible degrees of freedom. The properties of the free elastic modes of the non-rotating rotor and their consequences upon the necessary spatial integrals in the FFRF have been studied in detail by our group, and the vanishing of certain spatial integrals has been proved analytically. The free modes themselves and their non-vanishing spatial integrals can be obtained easily from a Finite Element pre-processing of the elastic rotor body, even for rotors with a complex shape and a strongly inhomogeneous material distribution. The non-linear equations of motion of the rotor are obtained afterwards in analytic form by means of symbolic computation, using the equations of Lagrange, see e.g. Shabana [4]. Particularly, we have used the commercial code Mathematica [5] to perform the symbolic computations. This formulation leads to a reduced model that remains valid also when the rigid-body degrees of freedom and their time-derivatives are not small. The model consists of a non-linear set of ordinary differential equations, and is denoted as fully non-linear model in the following. It can be integrated numerically using a suitable implicit code. In our work, the high-order integration code HOTINT [6] was utilized. It is noted that the above-mentioned pre-processing, before the time-integration, leads to a significant improvement of computation time in comparison to conventional multibody dynamics software. Details about the derivation of the fully non-linear model can be found in [7], where a successful comparison of the solutions of our approach with solutions of the commercial multibody software package ADAMS [3] was presented. An elastic rotor with unbalance forces, accelerated by a constant driving torque and having linear visco-elastic bearings, but no internal damping, the bearings being attached to the ground, was treated in [7]. The effects of different values of unbalance masses with respect to the resonance behavior of the elastic rotor were also studied in [7], demonstrating that above a certain critical unbalance mass, the bending resonance region cannot be surpassed, such that the axial rotation speed no longer can be increased, an effect which is accompanied by an increased vibration in the elastic degrees-of-freedom.

2 Consistent linearization of the fully non-linear rotor model

In [8], we have extended the formulation of [7] by first introducing modal damping for the flexible degrees-of-freedom, such that the effects of internal damping can be studied also, which may lead to an instability, see Gasch et al. [2] for a linear formulation. Moreover, a consistent linearization in the form of an expansion of the rigid-body tilting angles into a Maclaurin series has been presented in [8]. This is explained subsequently in more detail.

The center of mass of the co-rotating rigid rotor is used as the origin of a co-moving Cartesian frame, the axes of which are chosen as principle axes of inertia of the co-rotating rigid rotor. In order to describe the rotation of the co-moving Cartesian frame relative to the space-fixed frame, Bryant angles are used, see e.g. [9]. These angles are put into the column matrix

$$\theta = \left(\alpha\ \beta\ \gamma\right)^T \tag{1}$$

α and β are the so-called tilting angles, while γ is the rotation about the rotor axis. First, we rotate with the angle γ about the rotor axis, then with α and β about the transverse axes.

The angular velocity $\bar{\omega}$ of the co-rotating body in the co-rotating frame then can be expressed as

$$\bar{\omega} = \begin{pmatrix} 0 \\ \dot{\beta} \\ 0 \end{pmatrix} + A_\beta \begin{pmatrix} \dot{\alpha} \\ 0 \\ 0 \end{pmatrix} + A_\beta A_\alpha \begin{pmatrix} 0 \\ 0 \\ \dot{\gamma} \end{pmatrix} = \underbrace{\begin{pmatrix} \cos\beta & 0 & -\cos\alpha\sin\beta \\ 0 & 1 & \sin\alpha \\ \sin\beta & 0 & \cos\alpha\cos\beta \end{pmatrix}}_{\bar{G}} \underbrace{\begin{pmatrix} \dot{\alpha} \\ \dot{\beta} \\ \dot{\gamma} \end{pmatrix}}_{\dot{\theta}} \tag{2}$$

A_α and A_β denote elementary rotation matrices, well-known from the literature, see e.g. Shabana [4]. An overbar indicates the description in the co-rotating frame. Obviously, the tilting angles α and β can be assumed to be small in many applications.

In order to obtain a further model reduction, a partial linearization of the fully nonlinear system has been suggested in [8] which results in simplified equations and a further reduced computation time. As mentioned above, the fully non-linear model has been written in the form of the equations of Lagrange. The latter can be derived as a consequence of the fundamental law of dynamics, and their integrated forms, the relations of balance of momentum and angular momentum, see e.g. Ziegler [10].

The three rotational equations of Lagrange can be transformed such that they correspond to the relation of balance of angular momentum. Expressed in the co-moving Cartesian frame, this transformation can be formulated as

$$f = \left(A_\beta A_\alpha\right)^T \left(\bar{G}^{\mathrm{T}}\right)^{-1} \left(\frac{d}{dt}\left(\frac{\partial T}{\partial \dot{\theta}}\right) - \frac{\partial T}{\partial \theta} - F_\theta\right)^T = 0 \tag{3}$$

The kinetic energy is denoted by T, and F_θ is the generalized force for the rotational coordinates.

The partial linearization now is performed for the transformed expression given in (3), using a series expansion into a Maclaurin series. In order to fix ideas, and to obtain a connection to the classical formulations of linear rotordynamics, this procedure is explained for the rotational coordinates only in the following, writing

$$\begin{aligned} f \approx \hat{f}(\theta,\dot{\theta},\ddot{\theta}) = f\ |_{\alpha_0,\dots} &+ \tfrac{\partial f}{\partial \alpha}\ |_{\alpha_0,\dots}\ \alpha + \tfrac{\partial f}{\partial \beta}\ |_{\alpha_0,\dots}\ \beta + \\ &+ \tfrac{\partial f}{\partial \dot{\alpha}}\ |_{\alpha_0,\dots}\ \dot{\alpha} + \tfrac{\partial f}{\partial \dot{\beta}}\ |_{\alpha_0,\dots}\ \dot{\beta} + \\ &+ \tfrac{\partial f}{\partial \ddot{\alpha}}\ |_{\alpha_0,\dots}\ \ddot{\alpha} + \tfrac{\partial f}{\partial \ddot{\beta}}\ |_{\alpha_0,\dots}\ \ddot{\beta} \end{aligned} \tag{4}$$

For a rotor that is mounted on the ground, we may set $\alpha_0 = \beta_0 = \dot{\alpha}_0 = \dot{\beta}_0 = \ddot{\alpha}_0 = \ddot{\beta}_0 = 0$ in (4). Partial linearization of different types can be performed in (4). The axial rotational quantities γ, $\dot{\gamma}$ and $\ddot{\gamma}$ can not be considered as small in general. A linearization of the tilting angles α and β only, according to the first line in (4), yields for the rigid-body rotation portions

$$\begin{pmatrix} I_1 & 0 & 0 \\ 0 & I_2 & 0 \\ 0 & 0 & I_3 \end{pmatrix} \begin{pmatrix} \ddot{\alpha} \\ \ddot{\beta} \\ \ddot{\gamma} \end{pmatrix} + \dot{\gamma} \begin{pmatrix} 0 & I_3 - I_1 - I_2 & 0 \\ -(I_3 - I_1 - I_2) & 0 & 0 \\ 0 & 0 & 0 \end{pmatrix} \begin{pmatrix} \dot{\alpha} \\ \dot{\beta} \\ \dot{\gamma} \end{pmatrix}$$

$$
+ \dot{\gamma}^2 \begin{pmatrix} I_3 - I_2 & 0 & 0 \\ 0 & I_3 - I_1 & 0 \\ 0 & 0 & 0 \end{pmatrix} \begin{pmatrix} \alpha \\ \beta \\ \gamma \end{pmatrix} + \ddot{\gamma} \begin{pmatrix} 0 & I_3 - I_1 & 0 \\ -(I_3 - I_2) & 0 & 0 \\ 0 & 0 & 0 \end{pmatrix} \begin{pmatrix} \alpha \\ \beta \\ \gamma \end{pmatrix}
$$

$$
+ \underbrace{\begin{pmatrix} (I_3 - I_1)\, 2\beta\, \dot{\alpha}\, \dot{\beta} \\ (I_1 - I_3)\; \beta\, \dot{\alpha}^2 - (I_3 - I_1 + I_2)\, \alpha\, \dot{\alpha}\, \beta \\ 2\,(I_2 - I_3)\, \alpha\, \dot{\alpha}\, \dot{\gamma} + 2\;(I_1 - I_3)\, \beta\, \dot{\beta}\, \dot{\gamma} + (I_3 - I_1 + I_2)\, \dot{\alpha}\, \dot{\beta} \end{pmatrix}}_{a)}
$$

$$
+ \underbrace{\begin{pmatrix} 0 \\ 0 \\ (I_3 - I_1)\, \beta\, \ddot{\alpha} + I_2\, \alpha\, \ddot{\beta} \end{pmatrix}}_{b)} = \hat{F}_\theta \tag{5}
$$

In classical rotordynamic formulations, the tilting angles α, β as well as their time derivatives $\dot{\alpha}$, $\dot{\beta}$, $\ddot{\alpha}$, $\ddot{\beta}$ are considered to be small, see e.g. [1] and [2]. Thus, the term a) in (5) vanishes, if a linearization in $\dot{\alpha}$, $\dot{\beta}$ is performed, taking into account also the second line in (4). The same is valid for the term (b) according to a linearization in $\ddot{\alpha}$, $\ddot{\beta}$. In the next section it is demonstrated, however, that for a high angular rotation speed the terms (a) and (b) in (5) should not be neglected. In our formulation, we have performed partial linearizations not only with respect to the rotational coordinates, but also concerning the translational and flexible coordinates. This model is denoted as linearized model type 2, including terms analogous to the terms (a) and (b) in (5). Neglecting those terms leads to a further reduced model denoted as linearized model type 1, which for the rigid rotor corresponds to the models of classical rotordynamics presented by Krämer [1] and Gasch [2].

3 Numerical examples

In the following, numerical results stemming from the fully nonlinear and the linearized models type 1 and type 2, respectively, will be compared for various run-up simulations. Figure 1 shows a finite-element model of the suspended rotor modeled in ANSYS as well as the physical properties of the rotor and the parameters of the visco-elastic bearings. A turbine is mounted right-hand, whereas a compressor wheel is located on the opposite side. The rotor is accelerated from rest by a constant axial torque of 0.08 Nm, which is realized by four single forces acting in circumferential points at the outer radius of the turbine. An unbalance mass m_u is added onto a point on the outer radius of the center disc. No additional external forces are applied to the system. For all computations only the first two linear elastic modes are used for the Ritz approximation modeling the behavior of the flexible rotor. It turned out that using only these two elastic modes yields a sufficiently high accuracy. Internal damping of the free elastic modes is introduced via the

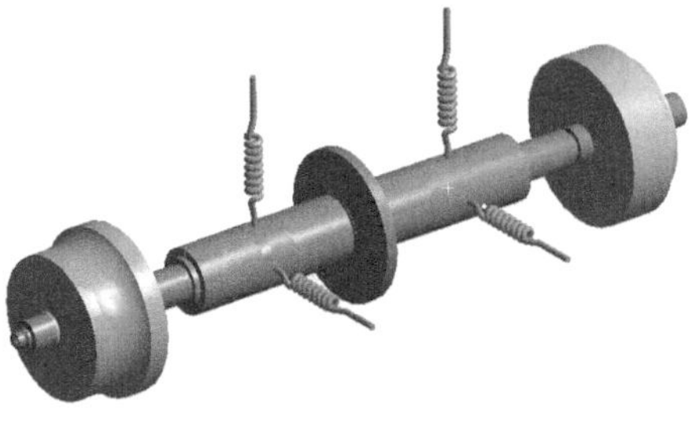

rotor properties		
total mass	m	0.19 kg
principal moments of inertia	I_1	3.5e−4 kgm²
	I_2	3.5e−4 kgm²
	I_3	1e−5 kgm²
visco-elastic bearings		
horizontal stiffness	k_x	1e4 N/m
vertical stiffness	k_y	1e5 N/m
axial stiffness	k_z	1e6 N/m
damping $d = d_x = d_y = d_z$	d	10 Ns/m

Fig. 1 Finite-element model of suspended rotor with mass and bearing properties

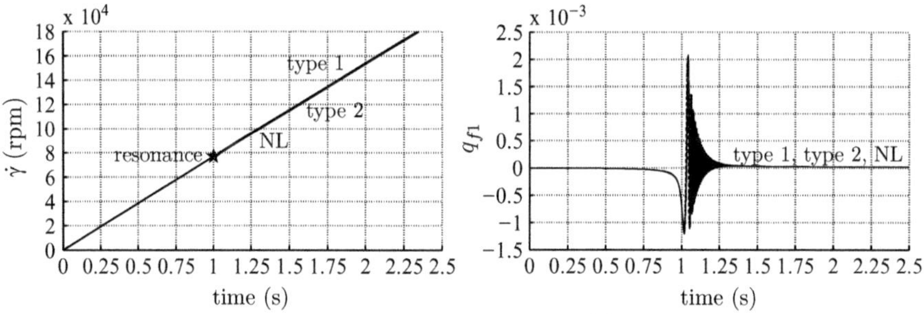

(a) rotor speed $\dot{\gamma}$

(b) flexible coordinate q_{f1}

Fig. 2 Run-up simulation $m_u = 1.1$ gmm, $D = 0$

Lehr damping ratio D. In the following simulations different phenomena regarding resonance and instability are studied for various values of unbalance mass m_u and damping ratio D. Figure 2 shows a run-up simulation where the rotational speed $\dot{\gamma}$ versus time is plotted in Fig. 2a and the first flexible coordinate q_{f1} is depicted in Fig. 2b. Using a small value of unbalance mass $m_u = 1.1$ gmm and vanishing modal damping $D = 0$ the rotational speed is increasing in a nearly linear way and the critical speed is passed generating small bending vibrations only, see Fig. 2b. The linearized models type 1 and type 2 successfully reproduce the behavior of the nonlinear model. For a higher value of unbalance mass $m_u = 5.5$ gmm and $D = 0$, flattening can be observed at the critical speed, see Fig. 3a. The flattening occurs due to resonance effects in the vicinity of the first bending eigenmode of the rotor. Type 1 fails and does not show the expected resonance phenomenon and passes the critical speed. Despite type 2 provides flattening, however the flexible coordinate does not coincide with the nonlinear result, see Fig. 3b.

In the sequel, a small unbalance mass $m_u = 1.1$ gmm is used. The results in Fig. 4 yield no noticeable collapse of rotor speed within the critical speed due to a resonance phenomenon, at 1s. However, the small modal damping $D = 0.005$ leads to an instability at higher rotor speed.

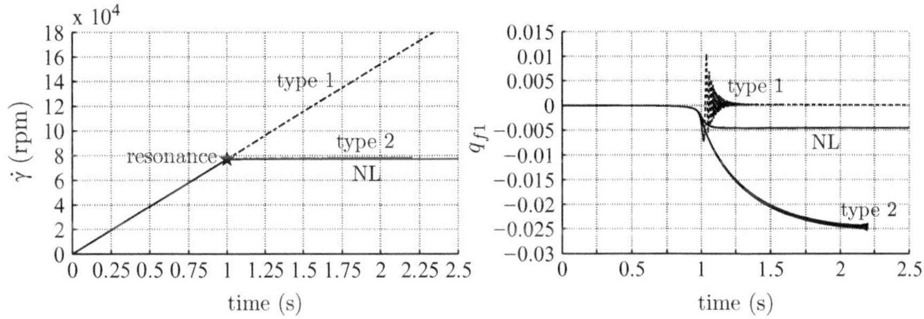

(a) rotor speed $\dot{\gamma}$ (b) flexible coordinate q_{f1}

Fig. 3 Run-up simulation $m_u = 5.5\,\text{gmm}$, $D = 0$

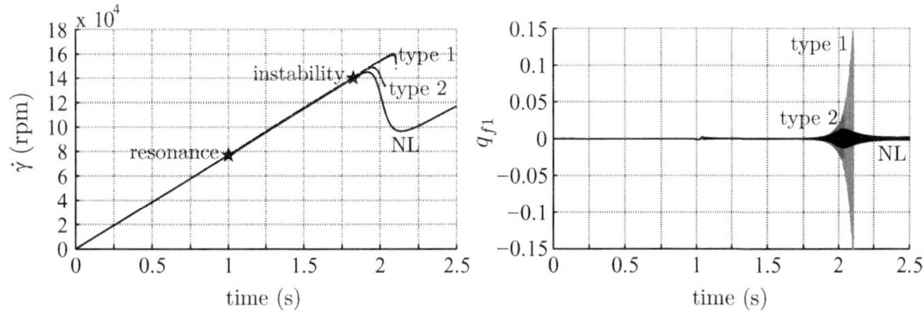

(a) rotor speed $\dot{\gamma}$ (b) flexible coordinate q_{f1}

Fig. 4 Run-up simulation $m_u = 1.1\,\text{gmm}$, $D = 0.005$

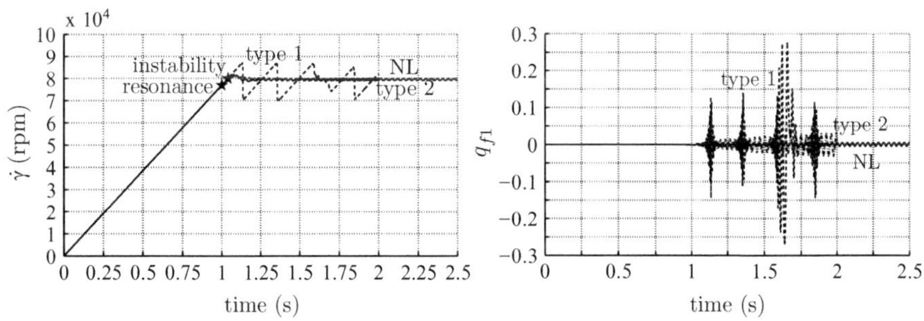

(a) rotor speed $\dot{\gamma}$ (b) flexible coordinate q_{f1}

Fig. 5 Run-up simulation $m_u = 1.1\,\text{gmm}$, $D = 0.1$

Despite the linearized models type 1 and type 2 show this instability phenomenon at higher speed, the solutions are not representing the nonlinear behavior of the rotor at all, since the elastic coordinates become unbounded. Increasing modal damping up to $D = 0.1$ pushes down the stability bound into the vicinity of the critical speed at resonance, see Fig. 5. Again, the linearized model type 1 totally fails. Type 2 shows the effect of instability, but the flexible coordinate is not bounded as in the nonlinear case.

An important result of this study is that the time-derivatives of the tilting angles and the flexible coordinates must not be considered as small for high-speed rotors especially for computing resonance and stability phenomena.

Acknowledgements Support of the present work in the framework of the peer-reviewed Austrian Center of Competence in Mechatronics (ACCM) is gratefully acknowledged.

References

1. Krämer, E.: Dynamics of Rotors and Foundations. Springer, Berlin (1993)
2. Gasch, R., Nordmann, R., Pfützner, H.: Rotordynamik. Springer, Berlin (2002)
3. ADAMS: www.mscsoftware.com/products/adams.cfm.
4. Shabana, A.A.: Dynamics of Multibody Systems, 3rd edn. Cambridge University Press, Cambridge (2005)
5. Mathematica. www.wolfram.com
6. Gerstmayr, J., Stangl, M.: High-order implicit Runge-Kutta methods for discontinuous mechatronical systems. Proceedings of XXXII Summer School APM, Russian Academy of Sciences, St. Petersburg, Russia, pp. 162–169, 2004, see also tmech.mechatronik.unilinz.ac.at/staff/gerstmayr/hotint.html
7. Irschik, H., Nader, M., Stangl, M., von Garssen, H.-G.: A floating-frame-of-reference formulation for deformable rotors using the properties of free elastic vibration modes. Proceedings of the ASME 2009 International Design Engineering Technical Conferences & Computers and Information Conference IDETC/CIE 2009, San Diego, California, USA, 2009, ASME-Paper No. DETC2009-86660
8. Nader, M., Irschik, H., Stangl, M., von Garssen, H.-G.: Nonlinear vibrations of flexible high-speed rotors supported by visco-elastic bearings, CD-ROM. Proceedings of the 8th IFToMM International Conference on Rotor Dynamics, Seoul, Korea, 2010
9. Geradin, M., Cardona, A.: Flexible Multibody Dynamics. Wiley, Chichester (2001)
10. Ziegler, F.: Mechanics of Solids and Fluids. Springer, Wien, NewYork, Corr. 2nd printing (1998)

Monitoring of Structural Deformations/ Vibrations Superposed upon Finite Pre-deformations

Michael Krommer and Yury Vetyukov

Abstract The signal of a continuously distributed strain-type sensor is proportional to the weighted integral of the local strains. Choosing the weight functions as a solution to a particular auxiliary problem of statics, we design a sensor, whose signal in the geometrically linear setting equals to a desired kinematic entity. In the present paper we study the design of such sensors in a geometrically nonlinear regime. In particular, we first discuss the implications of the type of the Lagrangian strain tensor, which is integrated by the sensor, on the problem of the design of the sensor. Then, we formulate the method of designing the sensor for the measurement of a kinematic entity in configurations, which remain in vicinity of a known largely pre-deformed state. The application of the method to the dynamical behavior of a straight beam under the action of a large moment is considered as an example problem.

1 Introduction

The design of a smart structure is a highly multi-disciplinary task, which involves the modeling of the structure, the interrogation and communication of the structure with a controller by means of suitable sensing and actuation, the integration of the smart system in the structure and the implementation of the system. Sensors and actuators are responsible for the functioning of the communication between structure and controller. Sensors provide information about the state the structure is

M. Krommer (✉)
Institute for Technical Mechanics, Johannes Kepler University Linz, Altenbergerstr. 69, A-4040 Linz, Austria
e-mail: michael.krommer@jku.at

Y. Vetyukov
Linz Center of Mechatronics GmbH, Altenbergerstr. 69, A-4040 Linz, Austria
e-mail: yury.vetyukov@lcm.at

H. Irschik et al. (eds.), *Advanced Dynamics and Model-Based Control of Structures and Machines*, DOI 10.1007/978-3-7091-0797-3_16,

in; this information has to be interpreted and properly processed by the controller to provide the actuator with information about what to do. For a discussion of frontiers in sensors/sensor systems see Glaser et al. [1]. In typical continuous systems a crucial point is the spatial distribution of sensors to obtain proper information as well as to perform distributed control of continua [2]. Finding these distributions for geometrically nonlinear problems is the main topic of the present paper.

In particular, we consider continuously distributed (or simply continuous) strain-type sensors. Those produce a signal that represents a weighted integral over the strain a body is suffering, in contrast to discrete strain-type sensors, whose signal is a sum of local strain values.

In the geometrically linear regime continuous sensors are well studied: e.g. Lee and Moon [3] introduced the concept of shaping the electrode pattern of a piezoelectric layer to measure a specific modal amplitude of a vibrating plate; among the vast amount of literature concerning the application of modal sensors we mention only [4,5]. Other types of sensors have been reported in the literature; e.g. nilpotent [6,7] or displacement sensors [8].

Although there is a large amount of literature available discussing structural problems with embedded strain-type sensors and actuators in the geometrically nonlinear regime [9, 10], little effort has yet been undertaken to study the effect of geometrical nonlinearities upon the design of strain-type sensors, rather than the effect on the sensor signal of a pre-designed sensor itself. In the present paper, we seek, in extension to the linear case, to measure structural entities of a small motion superposed upon a finite pre-deformation. The present paper provides new aspects to the previous analysis of the sensor design problem for nonlinear rods [11] and shells [12].

2 Continuous strain-type sensors

We study the motion of a three-dimensional material body, which undergoes finite deformation, but without any rigid body degrees of freedom. The volume in the reference configuration is V_0 and the boundary is Ω_0. The latter is composed of two parts $\Omega_{0,u}$ with prescribed homogenous kinematical boundary conditions for the displacements $\boldsymbol{u} = \mathbf{0}$ and $\Omega_{0,t}$, at which tractions $\boldsymbol{t}_0$ per unit undeformed area are applied. Body forces $\boldsymbol{b}_0$ per unit undeformed volume are acting. The position vector in the reference configuration is $\boldsymbol{r}$ and the one in the actual configuration is $\boldsymbol{R}(\boldsymbol{r})$ with the deformation gradient tensor $\mathbf{F} = \nabla_0 \boldsymbol{R}^T$; $d\boldsymbol{R} = \mathbf{F} \cdot d\boldsymbol{r}$; $\boldsymbol{u} = \boldsymbol{R} - \boldsymbol{r}$. Here, ∇_0 is the invariant differential operator with respect to the reference configuration and $\nabla = \mathbf{F}^{-T} \cdot \nabla_0$ is the one with respect to the actual configuration. We consider a general Lagrangian strain tensor $\mathbf{E} = \mathbf{f}(\mathbf{U})$, in which $\mathbf{U} = (\mathbf{F}^T \cdot \mathbf{F})^{\frac{1}{2}}$ is the stretch tensor and the strain measure function $\mathbf{f}$ is sufficiently smooth.

2.1 *Sensor design in the vicinity of a given configuration*

The sensorsignal of a continuous strain-type sensor of a material body undergoing finite deformation is defined as

$$Y = Y[\mathbf{E}] = \int_{V_0} \mathbf{S}_s(\boldsymbol{r}) \cdot\cdot\, \mathbf{f}(\mathbf{U}) d V_0. \tag{1}$$

Here $\mathbf{S}_s$ is a so-called shape tensor that defines the continuous distribution of the strain-type sensor. The situation is graphically shown in the following figure.

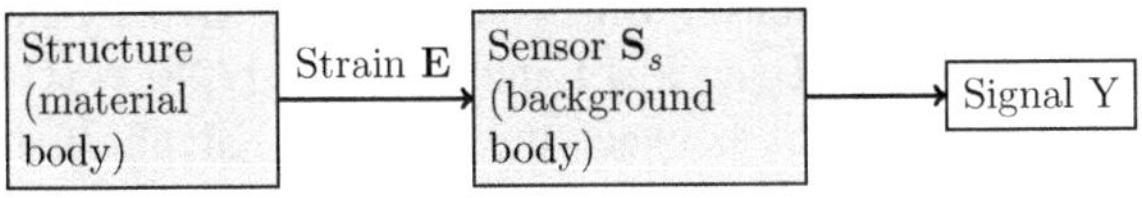

Here, the actual structure is a material body suffering a strain $\mathbf{E}$, which is transferred to the non-material sensing structure, and which only produces the signal as defined in (1).

As we have mentioned in the introduction the design problem for finite deformations is yet to be solved (if even possible!). Hence, we study the simpler problem of sensing small deformations relative to a given configuration, which is constituted by a volume V and a boundary $\Omega = \Omega_u + \Omega_t$. The position vector is $\boldsymbol{R}_u = \boldsymbol{R} + \boldsymbol{u}$, with $\boldsymbol{u}$ being the small additional displacement vector relative to the position in the design configuration. The new sensorsignal level is $Y_u = Y[\mathbf{E}_u]$ in which $\mathbf{E}_u \approx \mathbf{E} + \tilde{\mathbf{E}}$ follows from linearizing the general Lagrangian strain tensor in the vicinity of the design configuration. We further note, that the relation

$$\tilde{\mathbf{E}} = \frac{\partial \mathbf{f}}{\partial \mathbf{U}} \cdot\cdot\, 2\left(\mathbf{U} \boxtimes \mathbf{I} + \mathbf{I} \boxtimes \mathbf{U}\right)^{-1} \cdot\cdot \left(\mathbf{F}^T \cdot \boldsymbol{\varepsilon} \cdot \mathbf{F}\right) \tag{2}$$

holds for general Lagrangian strain tensors. The fourth rank tensor of the derivative $\partial \mathbf{f} / \partial \mathbf{U}$ is discussed in detail by [13]. In particular, it is positive definite, symmetric and invertible for any strain measure function [14]. The operation $\boxtimes$ has a meaning of the square tensor product, which was introduced by Del Piero in [15] and represents a combination of the common dyadic product of two second rank tensors with the subsequent transposition of the resulting fourth rank tensor with respect to the 2nd and 3rd indices. In (2) the small strain tensor in the vicinity of the design configuration is defined as $\boldsymbol{\varepsilon} = sym(\nabla \boldsymbol{u})$. A transformed sensorsignal, which is constituted as the deviation of the total sensorsignal Y_u from the sensorsignal Y that is measured for the body being in the design configuration can be written as

$$Y_u - Y = \int_V \boldsymbol{\sigma}_s \cdot\cdot\, \boldsymbol{\varepsilon}\, dV, \quad \boldsymbol{\sigma}_s = J^{-1} \mathbf{F} \cdot \bar{\mathbf{S}}_s \cdot \mathbf{F}^T, \tag{3}$$

$J = \det \mathbf{F}$ is the volume change. Here, we have introduced a transformed shape tensor $\bar{\mathbf{S}}_s$, which we may interpret as a 2nd Piola-Kirchhoff stress tensor, because of its transformation in (3). The original shape tensor is computed as:

$$\mathbf{S}_s(\mathbf{r}) = \frac{1}{2}\bar{\mathbf{S}}_s \cdot\cdot (\mathbf{U} \boxtimes \mathbf{I} + \mathbf{I} \boxtimes \mathbf{U}) \cdot\cdot \left(\frac{\partial \mathbf{f}}{\partial \mathbf{U}}\right)^{-1}. \tag{4}$$

Equation (4) can be easily applied to some common Lagrangian strain tensors. For the Green-Lagrange strain tensor we have $\mathbf{E} = \mathbf{f}(\mathbf{U}) = 1/2(\mathbf{U}^2 - \mathbf{I})$, such that $\partial\mathbf{f}/\partial\mathbf{U} = 1/2(\mathbf{U} \boxtimes \mathbf{I} + \mathbf{I} \boxtimes \mathbf{U})$ and $\mathbf{S}_s(\mathbf{r}) = \bar{\mathbf{S}}_s$; hence, the original shape tensor has the interpretation of a 2nd Piola-Kirchhoff stress tensor, which is work conjugate to the Green-Lagrange strain tensor. Likewise, using the Hencky strain tensor $\mathbf{E} = \mathbf{f}(\mathbf{U}) = \mathbf{U} - \mathbf{I}$ we have $\partial\mathbf{f}/\partial\mathbf{U} = \mathbf{I} \boxtimes \mathbf{I}$ and $\mathbf{S}_s(\mathbf{r}) = 1/2\bar{\mathbf{S}}_s \cdot\cdot (\mathbf{U} \boxtimes \mathbf{I} + \mathbf{I} \boxtimes \mathbf{U})$. This is the well known relation between the 2nd Piola-Kirchhoff stress tensor and the Biot stress tensor. In general (4) defines the work conjugate stress tensor for the Lagrangian strain tensor $\mathbf{E} = \mathbf{f}(\mathbf{U})$, if the transformed shape tensor $\bar{\mathbf{S}}_s$ is interpreted as a 2nd Piola-Kirchhoff stress tensor. This is only an interpretation, because the shape tensors (either $\bar{\mathbf{S}}_s$, $\mathbf{S}_s$ or $\boldsymbol{\sigma}_s$, if referred to the design configuration) are not real stress tensors. They only transform themselves between the reference configuration and the design configuration (which may not even be a configuration, that is actually occupied by the physical material body) according to rules valid for transformation of different stress tensors.

It is important to notice, that the deviation of the total sensorsignal Y_u from the sensorsignal Y that is measured for the body being in the design configuration, is independent from the special choice of the strain tensor $\mathbf{E} = \mathbf{f}(\mathbf{U})$. Only the sensorsignal in the design configuration depends on this latter choice, which is strictly correlated to the physical strain-type sensing mechanism that the material exhibits. In order to design sensors for the measurement of interesting structural (kinematical) entities in the vicinity of a design configuration, three main steps must be executed:

- The design configuration $\boldsymbol{R}(\boldsymbol{r})$ is either chosen or it is determined as a solution of a geometrically nonlinear problem.
- An auxiliary problem of statics for the design configuration is solved in a linear setting, producing $\boldsymbol{\sigma}_s$.
- The actual shape tensor for the reference configuration $\mathbf{S}_s$ is computed from $\bar{\mathbf{S}}_s = J\,\mathbf{F}^{-1} \cdot \boldsymbol{\sigma}_s \cdot \mathbf{F}^{-T}$ using (4).

These steps were first formulated in [11] for the particular case of the Green-Lagrange strain tensor. In the following subsection we will be discussing the second step in detail. The significance of the first step will be discussed in "Example Problem: Pure Bending of a Cantilever", whereas a detailed study of the implications of the third step is left for future research.

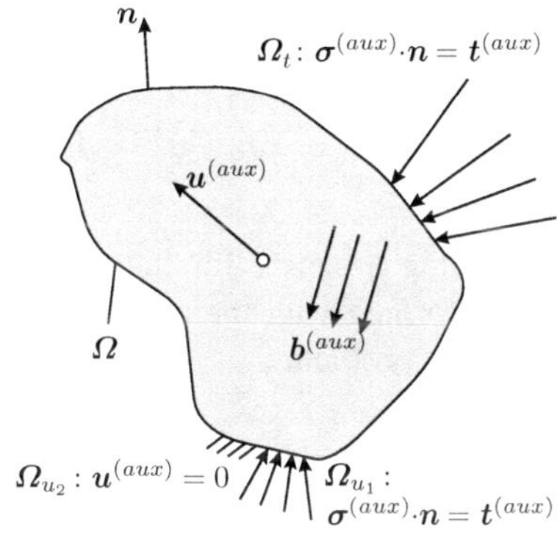

Fig. 1 Auxiliary quasi-static problem

2.2 *Auxiliary quasi-static problem*

As we have mentioned, an auxiliary problem of statics for the pre-deformed configuration must be solved in a linear setting to compute the sensing function. This problem can be related to the original problem by applying the principle of virtual work to the auxiliary problem of statics for the pre-deformed configuration. With the volume V and the boundary Ω in the pre-deformed configuration,

$$-\int_V \boldsymbol{\sigma}^{(aux)} \cdot\cdot \delta\boldsymbol{\varepsilon}\, dV + \int_V \boldsymbol{b}^{(aux)} \cdot \delta\boldsymbol{u}\, dV + \int_{\Omega_{u_1}+\Omega_t} \boldsymbol{t}^{(aux)} \cdot \delta\boldsymbol{u}\, d\Omega = 0, \qquad (5)$$

holds. Here, Ω_t is that part of the boundary, at which tractions are applied in the original problem, $\Omega_u = \Omega_{u_1} + \Omega_{u_2}$ has kinematical boundary conditions in the original problem; in the auxiliary problem a part Ω_{u_1} of Ω_u may as well have applied tractions (as long as no rigid body degrees of freedom exist in the auxiliary problem). The remaining part Ω_{u_2} has homogenous kinematical boundary conditions in the auxiliary problem, see Fig. 1. Furthermore, $\delta\boldsymbol{u}$, is any virtual displacement vector, which must be kinematically admissible, $\delta\boldsymbol{\varepsilon} = sym(\nabla\delta\boldsymbol{u})$ is the resulting kinematically admissible strain tensor and $\boldsymbol{\sigma}^{(aux)}$ is any statically admissible stress tensor. The latter satisfies the following problem of equilibrium:

$$\begin{aligned} V: &\quad \nabla\cdot\boldsymbol{\sigma}^{(aux)} + \boldsymbol{b}^{(aux)} = \boldsymbol{0}, \\ \Omega_{u_1} + \Omega_t: &\quad \boldsymbol{\sigma}^{(aux)} \cdot \boldsymbol{n} = \boldsymbol{t}^{(aux)}. \end{aligned} \qquad (6)$$

In order to relate this auxiliary problem to our original problem of small deformations superposed upon a pre-deformed configuration, the small additional displacement $\boldsymbol{u}$ relative to the position in the pre-deformed state (e.g. due to vibrations) is considered. Using $\boldsymbol{u}$ instead of $\delta\boldsymbol{u}$ in (5), after simple mathematical transformations [11] we can show that the sensor with $\boldsymbol{\sigma}_s = \boldsymbol{\sigma}^{(aux)}$ in (3) measures the work of the external auxiliary forces on the original small additional deformation:

$$Y_u - Y = \int_V \boldsymbol{b}^{(aux)} \cdot \boldsymbol{u}\, dV + \int_{\Omega_t} \boldsymbol{t}^{(aux)} \cdot \boldsymbol{u}\, d\Omega. \tag{7}$$

Designing a sensor for the measurement of a particular kinematical entity can be performed with the appropriate choice of the loadings $\boldsymbol{t}^{(aux)}$ and $\boldsymbol{b}^{(aux)}$ in the auxiliary problem.

3 Example problem: pure bending of a cantilever

We study the described sensor design methodology for the simple example of a cantilever with a single moment applied at the tip. For this simple example the solution for the sensor design problem can be found in a closed form.

A rod of the length $l = 1$ is clamped at one end and loaded by a moment M at the other one. A quasistatic solution of the problem is a circular arc with the curvature κ, which is proportional to M; the rod is bent into a full circle with the curvature $1/2\pi$ when the moment reaches a particular value M_1. We consider a piecewise linear time history of the loading:

$$M(t) = \begin{cases} M_1 t/t_1, & t < t_1 \\ M_1, & t \geq t_1. \end{cases} \tag{8}$$

The time range t_1 was chosen such that the dynamic effects would play a moderate role in the solution. Deformed configurations of the rod for three moments in time, obtained either by solving dynamical equations (dynamic) or static equations (quasistatic) are presented in Fig. 2. The time history of the vertical displacement of the end point of the rod is plotted in Fig. 3 for the dynamic, quasistatic and linear quasistatic solutions. The static displacement depends on the moment non-monotonously, and the sensor signal is proportional to it (as it is proportional to the curvature). Therefore a universal sensor, which would measure this displacement in the whole geometrically nonlinear range, cannot be designed.

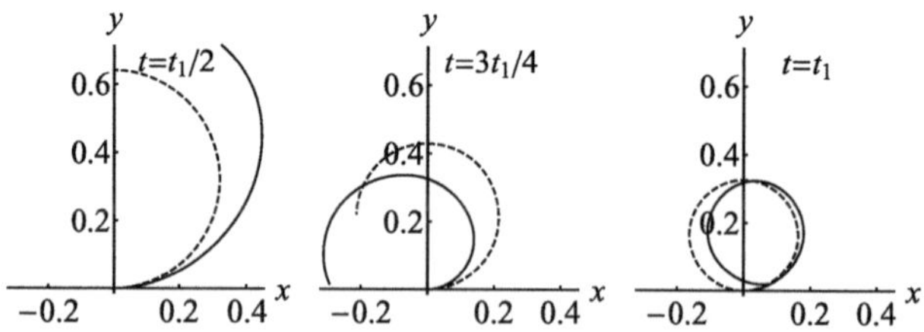

Fig. 2 Dynamic (*solid lines*) and quasistatic (*dashed lines*) configurations at different time instants

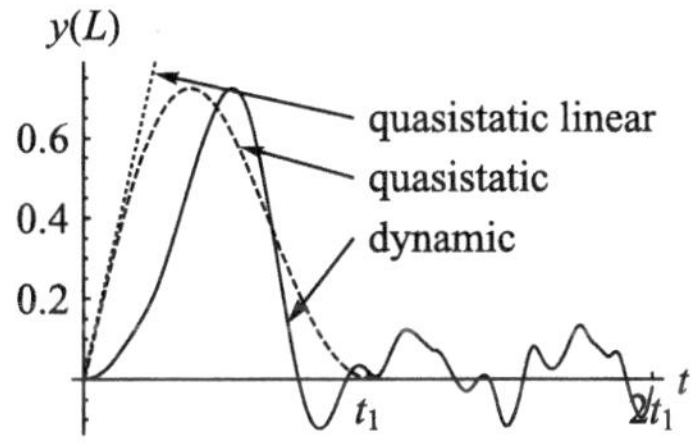

Fig. 3 Vertical coordinate of the tip in the quasistatic and dynamic solutions

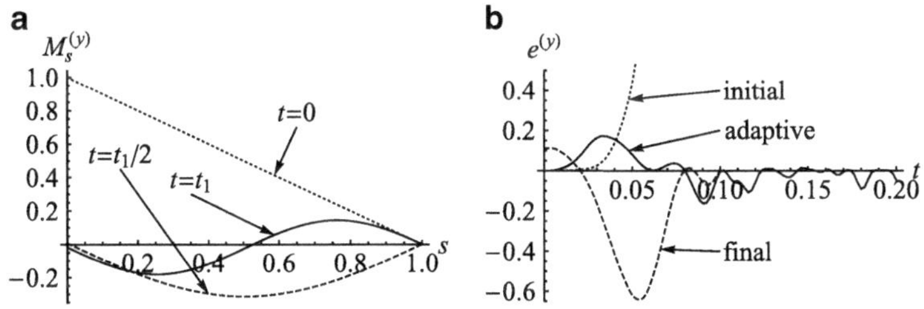

Fig. 4 Sensor shape functions at three quasistatic configurations (**a**) and error signals for three strategies of the sensor design (**b**)

Applying the strategy, presented in the theoretical part of the paper, we can design a sensor for small deviations from each quasistatically pre-deformed configuration:

$$Y = \int_0^l M_s^{(y)}(s)\kappa(s)\,ds. \tag{9}$$

$M_s^{(y)}(s)$ is a distribution of the bending moment in the appropriate problem of statics of the curved rod under the action of a unit tip force. Distributions of $M_s^{(y)}(s)$ for three different pre-deformed configurations are presented in Fig. 4a. We considered three sensor designs: for the initial (straight) configuration, for the final (circular) configuration, and adaptive, in which the used distribution of the sensor weight $M_s^{(y)}(s)$ was updated in time according to the present load value. Errors of the measurement of the vertical position in the simulated dynamic process are compared in Fig. 4b.

Acknowledgements Support of the authors from the Austrian Science Fund (FWF Translational project L441-N41 *Sensor Systems for Structural and Health Monitoring*) and from the Austrian Center of Competence in Mechatronics (ACCM) is gratefully acknowledged.

References

1. Glaser, S.D., Shoureshi, R.A., Pescovitz, D.: Frontiers in sensors and sensing systems. Smart Struct. Syst. **1**(1), 103–120 (2005)
2. Gabbert, U., Tzou, H.S.: Preface: Proceedings of IUTAM-Symposium on Smart Structures and Structronic Systems. Magdeburg, Germany, September 2000, 2001.
3. Lee, C.-K., Moon, F.-C.: Modal sensors/actuators. J. Appl. Mech. **57**, 434–441 (1990)
4. Tzou, H.S., Hollkamp, J.J.: Collocated independant modal control with self-sensing orthogonal piezoelectric actuators (theory and experiment). Smart Mater. Struct. **3**, 277–284 (1994)
5. Sun, D., Tong, L.: Modal control of smart shells by optimized discretely distributed piezoelectric transducers. Int. J. Solids Struct. **38**(18), 3281–3299 (2001)
6. Miu, D.K.: Mechatronics: Electromechanics and Contromechanics. Springer, New York (1992)
7. Irschik, H., Krommer, M., Belyaev, A.K., Schlacher, K.: Shaping of piezoelectric sensors/actuators for vibrations of slender beams: coupled theory and inappropriate shape functions. J. Intell. Mater. Syst. Struct. **9**, 546–554 (1999)
8. Irschik, H., Krommer, M., Pichler, U.: Shaping of distributed piezoelectric sensors for flexural vibrations of smart beams. In: Proceedings of SPIE's 6th Annual International Symposium on Smart Structures and Materials. Newport Beach, CA, March 1999
9. Tylikowski, A., Frischmuth, K.: Stability and stabilization of circular plate parametric vibrations. Int. J. Solids Struct. **40**(19), 5187–5196 (2003)
10. Cheng, J., Wang, B., Du, S.Y.: A theoretical analysis of piezoelectric/composite anisotropic laminate with larger-amplitude deflection effect, Part I: Fundamental equations. Int. J. Solids Struct. **42**(24–25), 6166–6180 (2005)
11. Vetyukov, Yu., Krommer, M.: Optimal continuous strain-type sensors for finite deformations of shell structures. Mech. Adv. Mater. Struct. **18** (2011)
12. Krommer, M., Vetyukov, Yu.: Adaptive sensing of kinematic entities in the vicinity of a time-dependent geometrically nonlinear pre-deformed state. Int. J. Solids Struct. **46**, 3313–3320 (2009)
13. Norris, A.N.: Higher derivatives and the inverse derivative of a tensor-valued function of a tensor. Q. Appl. Math. **66**, 725–741 (2008)
14. Hill, R.: Aspects of invariance in solid mechanics. Adv. Appl. Mech. **18**, 1–75 (1978)
15. Del Piero, G.: Some properties of the set of fourth-order tensors with application to elasticity. J. Elasticity **3**, 245–261 (1979)

Electromechanical and Mathematical Models of Salient-Pole Synchronous Motors

G.A. Leonov and N.V. Kondrat'eva

Abstract Two electromechanical models of salient-pole synchronous motor: a classical two-pole model and four-pole model are considered in terms of the coordinates rigidly connected to rotating magnetic field. The four-pole model describes electromagnetic processes in salient-pole motor more completely than two-pole model. The four-pole model is considered for different commutations in excitation system of synchronous motor. The comparison of the models of rotor of salient-pole synchronous motor allows one to make the following conclusion: for the four-pole model of rotor the effect of excitation windings on damping moment is greater than for the two-pole one.

1 Introduction

In 1888 Tesla and Ferraris invented a rotating magnetic field generated by alternate current in stationary windings of stator of electrical machine. Till now this phenomenon is a base of constructing alternating current machines: the synchronous and asynchronous ones [2, 8–10, 12, 14, 17, 20].

We shall show that derivation of differential equations describing such electrical machines can be substantially simplified. For this purpose the rotating system of coordinates rigidly connected to the rotating magnetic field is introduced. In this coordinate system a steady operating mode for synchronous machine under constant loads is a stationary state of the rotor.

The introduction of coordinates rigidly connected to rotating magnetic field allows obtaining more accurate mathematical models of synchronous machines.

G.A. Leonov (✉) · N.V. Kondrat'eva
Department of Mathematics and Mechanics, St. Petersburg State University,
Universitetskaya av., 28, Petrodvorets, 198504 St. Petersburg, Russia
e-mail: leonov@math.spbu.ru; knat0202@mail.ru

H. Irschik et al. (eds.), *Advanced Dynamics and Model-Based Control of Structures and Machines*, DOI 10.1007/978-3-7091-0797-3_17,

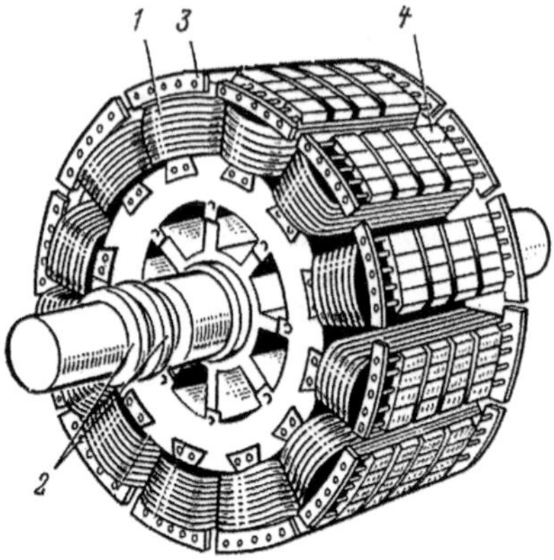

Fig. 1 Rotor of the salient-pole synchronous motor. 1 is excitation winding, 2 is collector rings, 3 is damping winding, 4 is pole

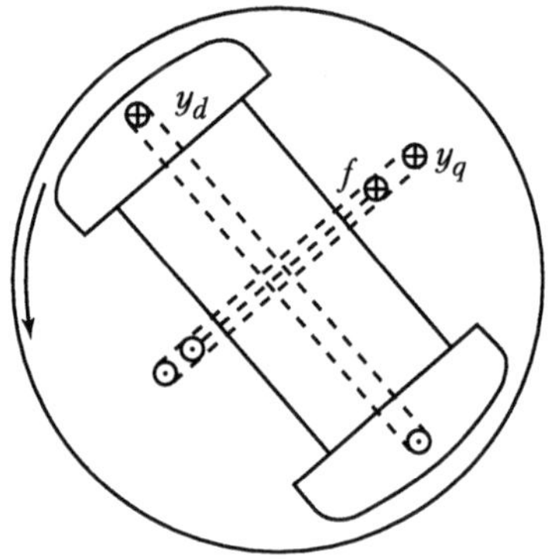

Fig. 2 Schematic of two-pole rotor. f is excitation winding, y_q and y_d are damping windings

It is clear that any mathematical model of electrical machine is a certain idealization. While constructing such a model it is useful to make an intermediate step and consider the electromechanical model of electrical machine. Substantially, such an intermediate idealization occurs, explicitly or implicitly, in all works devoted to mathematical description of synchronous electrical machines.

Synchronous machines obey the inversability principle and, therefore, they can operate as either generator or motor. The principle of inversability allows one to conclude that mathematical models of synchronous machines operating in the mode of electric energy generation preserve the same structure as synchronous electric motors to be considered below. In what follows we consider synchronous machines operating as a motor.

The electromechanical models of synchronous motor with salient-pole rotor (Fig. 1) are studied. We consider two-pole and four-pole electromechanical models, shown in Figs. 2 and 3, respectively, [2, 8–12, 14, 17, 20, 21].

For the models of Figs. 2 and 3 a further idealization depicted in Figs. 4 and 5, is used. In Fig. 4 one sees a schematic drawing of two-pole model of rotor, a scheme of three winds: a wind of excitation winding with the current $i(t)$ and two orthogonal

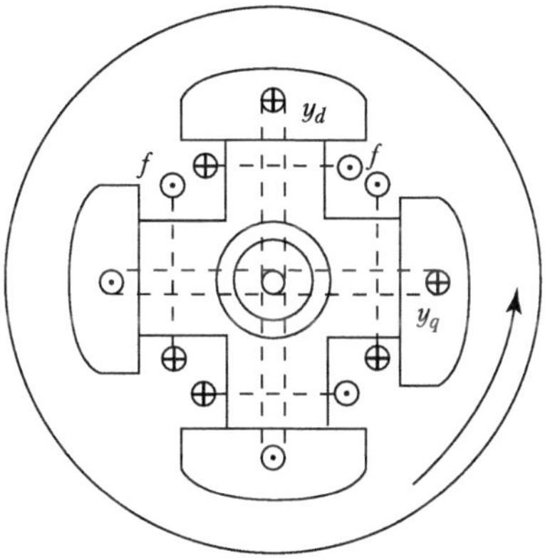

Fig. 3 Schematic of four-pole rotor. excitation windings (f) and damping windings (y_d, y_q) are shown

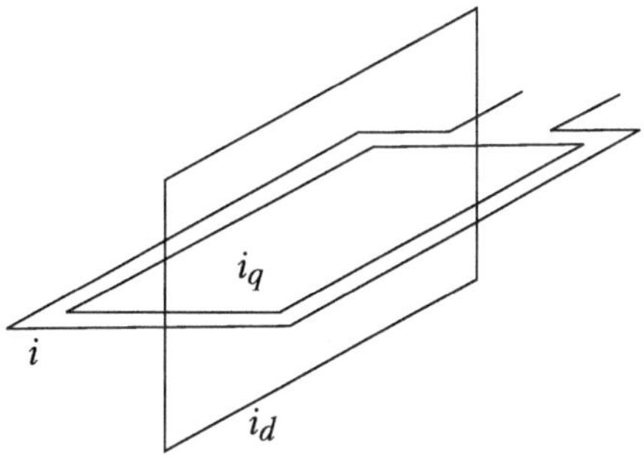

Fig. 4 Scheme of three windings

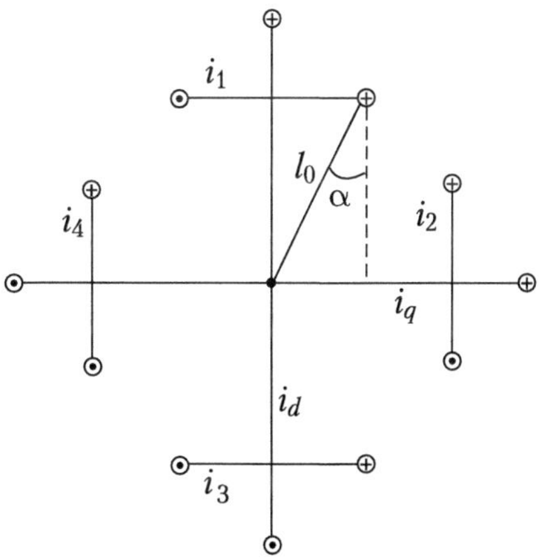

Fig. 5 Scheme of six windings. Two orthogonal pairs of parallel winds of excitation winding and a pair of orthogonal damping windings (a pair with the currents $i_1(t)$ and $i_2(t)$, a pair with the currents $i_3(t)$ and $i_4(t)$, and a pair with the currents $i_q(t)$ and $i_d(t)$) are shown

winds of damping windings with currents $i_q(t)$ and $i_d(t)$, respectively. Figure 3 displays a scheme of four-pole model of rotor with six windings. This scheme consists of two orthogonal pairs of parallel winds of excitation winding (a pair with the currents $i_1(t)$ and $i_3(t)$ and a pair with the currents $i_2(t)$ and $i_4(t)$) and a centered orthogonal pair of damping windings with the currents $i_q(t)$ and $i_d(t)$.

The classical laws of mechanics and electrical engineering make it possible to proceed to mathematical modelling from the electromechanical models.

Let us notice that the well-known classical model of synchronous motor, the model of Park–Gorev, is a two-pole model shown in Fig. 2, cf. [9, 16]. In classical investigations the rotor of synchronous motor without damping windings experiences weakly damped periodic oscillations in the rotating coordinate system and one observes a long transition to operating mode mentioned above. Following the analysis of four-pole model below we obtain that for some commutations the very disposition of excitation winding on salient-pole rotor generates considerable damping moment which rapidly suppresses the rotor oscillations. Thus, after a transient process a process of synchronization occurs: a rotor rotates with the frequency of magnetic field.

In four-pole model there exist three types of commutations in the system of excitation of synchronous motor, namely three different ways of connection of excitation winding to a constant voltage source:

1. The commutation of type I (four excitation currents)
2. The commutation of type II (two excitation currents)
3. The commutation of type III (one excitation current)

All the equations derived here are obtained by means of the general approach. It implies introducing the system of coordinates rigidly connected to rotating magnetic field and considering motion of electromechanical model of synchronous motor in this system of coordinates. This consideration is transparent and simplifies the derivation of equations under the above assumptions.

2 Two-pole model of rotor

We proceed from the electromechanical models of synchronous motors to their mathematical description. Let us study an electromechanical model of two-pole rotor. We consider a scheme of three windings shown in Fig. 4 and begin with study of the motion of windings in the rotating coordinate system rigidly connected to the magnetic field vector **B**. Let us introduce the following parameters of the windings: L, L_q, L_d are inductivities, R, R_q, R_d are armature resistances, and S, S_q, S_d are areas of separate winds of excitation winding and damping windings, respectively. Further n, n_q, n_d denote the numbers of winds in excitation winding and damping windings, respectively, and θ stands for the angle between the plane of wind of winding with the current $i(t)$ and a plane perpendicular to magnetic field vector.

In what follows the mutual induction of windings is neglected. By Ohm's law and the law of electromagnetic induction we obtain the following system of equations for the excitation current $i(t)$, damping currents $i_q(t)$,$i_d(t)$ and the angle θ [14]

$$L\frac{di(t)}{dt} + R \cdot i(t) = nSB \cdot \sin\theta(t) \cdot \dot{\theta}(t) + e, \tag{1}$$

$$L_q \cdot \frac{di_q(t)}{dt} + R_q \cdot i_q(t) = n_q S_q B \cdot \sin\theta(t) \cdot \dot{\theta}(t), \tag{2}$$

$$L_d \cdot \frac{di_d(t)}{dt} + R_d \cdot i_d(t) = n_d S_d B \cdot \cos\theta(t) \cdot \dot{\theta}(t). \tag{3}$$

$$I\ddot{\theta} = -B(nSi(t)\sin\theta + n_q S_q i_q(t)\sin\theta + n_d S_d i_d(t)\cos\theta) - M. \tag{4}$$

Neglecting inductivities of damping windings, i.e. assuming $L = L_q = L_d = 0$ in (1)–(3), system (1)–(4) can be reduced to third order system by nonsingular transformation of coordinates. This system is considered in the monographs [14–21] as equations for synchronous motor for some approximation of the damping windings. In [14] system (1)–(4) is reduced to a well-known fifth order system of equations of synchronous motor studied in [7].

Assuming $L = L_q = L_d = 0$ in (1)–(3) and replacing θ by $-\theta$ we obtain

$$\ddot{\theta} + f(\theta)\dot{\theta} + b\sin\theta = \gamma, \tag{5}$$

where function $f(\theta)$ and parameters b, γ are given by

$$f(\theta) = \left(\frac{(nSB)^2}{IR} + \frac{(n_q S_q B)^2}{IR_q}\right)\sin^2\theta + \frac{(n_d S_d B)^2}{IR_d}\cos^2\theta$$

$$b = \frac{nSBe}{IR}, \quad \gamma = \frac{M}{I}. \tag{6}$$

Note that (5) describes dynamics of synchronous motor in the simplest two-dimensional idealization [1, 3–6, 13, 15, 19]. In (5) $f(\theta)\dot{\theta}$ corresponds to damping moment, b defines the maximum of static characteristic of synchronous motor and γ corresponds to the external load moment. A particular case of (5) takes the form [3–6, 13]:

$$\ddot{\theta} + k(1 - d\cos 2\theta)\dot{\theta} + b\sin\theta = \gamma. \tag{7}$$

Equation (7) is known to govern the pendulum motion under constant force if $d = 0$. It was regarded first as the simplest model of synchronous machine in [5, 6, 15, 19].

We introduce the following notation:

$$k_e = (nSB)^2/IR, \quad k_q = (n_q S_q B)^2/IR_q, \quad k_d = (n_d S_d B)^2/IR_d.$$

Notice that parameters k_e, k_q, k_d show how the excitation winding and damping windings affect respectively the damping moment. For some relations between parameters k_e, k_q, k_d (5) reduces to (7).

Making use of (1)–(4) we can obtain a third order system of two-pole rotor without damping windings and (7) in which $k = k_e, d = 1$, cf. [11, 14, 18]. Thus, when the damping windings are absent a weak damping moment can be supplied by the excitation winding.

3 Four-pole model of rotor

Let us consider the electromechanical model of four-pole rotor. We address the scheme of six windings shown in Fig. 5 for different commutations in the excitation system. It can be shown that four-pole model of rotor is a generalization of two-pole model.

As before, we consider motion of windings in the rotating coordinate system rigidly connected to the magnetic field vector. We keep the previous denotations for parameters of the exiting windings. The parameters of damping windings are assumed to be coincident, i.e. $L_q = L_d = L_0$, $R_q = R_d = R_0$, $S_q = S_d = S_0$. Then $k_q = k_q = k_0$.

We study first the case of the commutation of type I, i.e. the case in which the excitation currents $i_1(t), i_2(t), i_3(t), i_4(t)$ are different. By Ohm's law and the electromagnetic induction we obtain the following system of equations

$$L\,\frac{di_1(t)}{dt} + Ri_1(t) = e + nSB\sin\theta(t)\cdot\dot{\theta}(t), \tag{8}$$

$$L\,\frac{di_2(t)}{dt} + Ri_2(t) = e + nSB\cos\theta(t)\cdot\dot{\theta}(t), \tag{9}$$

$$L\,\frac{di_3(t)}{dt} + Ri_3(t) = e + nSB\sin\theta(t)\cdot\dot{\theta}(t), \tag{10}$$

$$L\,\frac{di_4(t)}{dt} + Ri_4(t) = e + nSB\cos\theta(t)\cdot\dot{\theta}(t). \tag{11}$$

$$L_0\,\frac{di_q(t)}{dt} + R_0 i_q(t) = n_0 S_0 B\sin\theta(t)\cdot\dot{\theta}(t), \tag{12}$$

$$L_0\,\frac{di_d(t)}{dt} + R_0 i_d(t) = n_0 S_0 B\cos\theta(t)\cdot\dot{\theta}(t). \tag{13}$$

$$\begin{aligned} I\ddot{\theta} &= -nSB[(i_1(t)+i_3(t))\sin\theta + (i_2(t)+i_4(t))\cos\theta] \\ &\quad -n_0 S_0 B[i_q(t)\sin\theta + i_d(t)\cos\theta] - M. \end{aligned} \tag{14}$$

Here $\theta(t)$ is the angle between the planes of winds of the windings with currents $i_1(t), i_3(t)$ and the plane perpendicular to magnetic field vector; $i_1(t), i_2(t)$ are excitation currents and $i_q(t), i_d(t)$ are damping currents.

Consider further a scheme of six windings (Fig. 5) for the commutation of type II in two particular cases: in the case of the so-called "series" commutation when $i_1(t) = i_2(t) := i_1(t)$, $i_3(t) = i_4(t) := i_2(t)$ and in the case of the so-called "parallel" commutation when $i_1(t) = i_3(t) := i_1(t)$, $i_2(t) = i_4(t) := i_2(t)$. In the case of "series" commutation the dynamics of four-pole rotor with damping windings is governed by the following system of equations

$$L\,\frac{di_1(t)}{dt} + Ri_1(t) = e + 2nSB(\sin\theta(t) + \cos\theta(t))\cdot\dot{\theta}(t), \tag{15}$$

$$L\,\frac{di_2(t)}{dt} + Ri_2(t) = e + 2nSB(\sin\theta(t) + \cos\theta(t))\cdot\dot{\theta}(t), \tag{16}$$

$$L_0 \frac{di_q(t)}{dt} + R_0 i_q(t) = n_0 S_0 B \sin\theta(t) \cdot \dot{\theta}(t), \tag{17}$$

$$L_0 \frac{di_d(t)}{dt} + R_0 i_d(t) = n_0 S_0 B \cos\theta(t) \cdot \dot{\theta}(t), \tag{18}$$

$$I\ddot{\theta} = -nSB[(\sin\theta + \cos\theta)\,(i_1(t) + i_2(t))] - n_0 S_0 B[i_q(t)\sin\theta + i_d(t)\cos\theta] - M. \tag{19}$$

In the case of "parallel" commutation the system of equations is as follows

$$L\frac{di_1(t)}{dt} + Ri_1(t) = e + 2nSB\sin\theta(t) \cdot \dot{\theta}(t), \tag{20}$$

$$L\frac{di_2(t)}{dt} + Ri_2(t) = e + 2nSB\cos\theta(t) \cdot \dot{\theta}(t), \tag{21}$$

$$L_0 \frac{di_q(t)}{dt} + R_0 i_q(t) = n_0 S_0 B \sin\theta(t) \cdot \dot{\theta}(t), \tag{22}$$

$$L_0 \frac{di_d(t)}{dt} + R_0 i_d(t) = n_0 S_0 B \cos\theta(t) \cdot \dot{\theta}(t). \tag{23}$$

$$I\ddot{\theta} = -2nSB[i_1(t)\sin\theta + i_2(t)\cos\theta] - n_0 S_0 B[i_q(t)\sin\theta + i_d(t)\cos\theta] - M. \tag{24}$$

In systems (15–19) and (20–24) $i_1(t), i_2(t)$ are excitation currents and $i_q(t), i_d(t)$ are the damping currents.

If we set $i_1(t) = i_2(t) = i_3(t) = i_4(t) = i(t)$ in the system (8–14) then in the case of the commutation of type III (with one excitation current $i(t)$) we obtain the following system of equations for four-pole rotor with damping windings

$$L\frac{di(t)}{dt} + Ri(t) = e + 2nSB(\sin\theta(t) + \cos\theta(t)) \cdot \dot{\theta}(t), \tag{25}$$

$$L_0 \frac{di_q(t)}{dt} + R_0 i_q(t) = n_0 S_0 B \sin\theta(t) \cdot \dot{\theta}(t), \tag{26}$$

$$L_0 \frac{di_d(t)}{dt} + R_0 i_d(t) = n_0 S_0 B \cos\theta(t) \cdot \dot{\theta}(t). \tag{27}$$

$$I\ddot{\theta} = -2nSB(\sin\theta + \cos\theta)i(t) - n_0 S_0 B[i_q(t)\sin\theta + i_d(t)\cos\theta] - M. \tag{28}$$

Neglecting the inductivities of windings in (8–14), (15–19), (20–24), (25–28) (i.e. assuming $L = L_0 = 0$) and replacing θ by $-\theta - \pi/4$ we obtain (7). For the case of commutation of type I the coefficients of this equation are now given by the formulae

$$f(\theta) = 2k_e + k_0, \quad b = 2\sqrt{2}\ nSBe/IR, \quad \gamma = M/I, \quad d = 0. \tag{29}$$

In the case of commutations of types II and III the parameters b, γ are given by formulae (29). In the case of "parallel" commutation of type II the parameters k, d are defined by formulae $k = 4k_e + k_0$, $d = 0$ whereas in the cases of the "series" commutation of type II and the commutation of type III they are as follows $k = 4k_e + k_0$, $d = k_e/4(k_e + k_0)$.

Thus while studying the four-pole model of rotor we arrive at (7) with the parameters k (the coefficient of damping moment) and b (the maximum of static characteristic), their values being greater than those for the two-pole model of rotor. It should be noticed that the damping moment provides a stable operation of synchronous motor and the maximum of static characteristic defines the value of the limiting load for synchronous motor, cf. [1, 14, 21].

References

1. Barbashin, E.A., Tabueva, V.A.: Dynamical Systems with Cylindric Phase Space. Nauka, Moscow (1969)
2. Begamudre, R.D.: Electromechanical Energy Conversion with Dynamics of Machines. New Age International, New Delhi (2007)
3. Belyustina, L.N.: Stability of operating regime of salient-pole synchronous motor. Izv. AN SSSR OTN. **10**, 131–140 (1954)
4. Belyustina, L.N.: On an equation of theory of electrical machines. In: Sb. to the memory of A.A. Andronov, pp. 173–186. Isd. AN SSSR, Moscow (1955)
5. Coppel, W.A.: On the equation of a synchronous motor. Quart. J. Mech. Appl. Math. **12**(2), 242–256 (1959)
6. Edgerton, H.E., Fourmarier, P.: The pulling into step of a saleint-pole synchronous motor. AIEE Trans. **50**, 769–778 (1931)
7. Fagiuoli, E., Szegö, G.P.: Qualitative analysis by Modern methods of a stability problem in power-system analysis. J. Franklin Inst. **290**(2), 103–111 (1970)
8. Fitzgerald, A.E., Kingsley, C., Jr, Umans, S.D.: Electric Machinery. McGraw-Hill, New-York (1999)
9. Gorev, A.A.: Transient Processes of Synchronous Machine. Nauka, Leningrad (1985)
10. Gross, C.A.: Electric Machines. CRC Press, New-York (2006)
11. Halanay, A., Leonov, G.A., Rasvan, V.: From pendulum equation to an extended analysis of synchronous machines. Rend. Sem. Math. Univers. Politecn. Torino. **45**(2), 91–106 (1987)
12. Ivanov-Smolenskii, A.B.: Electrical Machines. Energiya, Moscow (1980)
13. Ku, Y.H.: Non-linear analysis of electromechanical problems. J. Franklin Inst. **255**(1), 9–32 (1953)
14. Leonov, G.A., Kondrat'eva, N.V.: Stability analysis of electric alternating current machines. Isd. St. Petersburg University, St. Petersburg (2009)
15. Lyon, W.V., Edgerton, H.E.: Transient torque – angle charateristics of synchronous mashines. AIEE Trans. **49**, 686–699 (1930)
16. Park, R.H.: Two-reaction theory of synchronous machines generalized method of analysis. AIEE Trans. **48**(1), 716–727 (1929)
17. Sarma, M.S.: Synchronous Machines: Their Theory, Stability, and Excitation Systems. Gordon and Breach Science, New-York (1979)
18. Szego, G.P., Olech, C., Cellina, A.: On the stability properties of a third order system. Ann. mat. pura ed appl. (**91**), 91–103 (1968)
19. Tricomi, F.: Integrazione di unequazione differenziale presentatasi in elettrotechnica. Annali della R. Scuola Normale Superiore di Pisa. **2**(2), 1–20 (1933)
20. Vazhnov, A.I.: Electrical Machines. Energiya, Leningrad (1969)
21. Yanko-Trinitskii, A.A.: New method for analysis of operation of synchronous motor for jump-like loads. GEI, Moscow-Leningrad (1958)

Optimization of Dynamic Characteristics of Electroviscoelastic Systems with External Electric Circuits

V.P. Matveenko, E.P. Kligman, M.A. Yurlov, and N.A. Yurlova

Abstract The paper presents solutions to the optimization problems considering the dynamic characteristics (resonance and dissipative properties) of electro-visco-elastic deformable systems, the behavior of which is described by the equations of linear elasticity and linear hereditary viscoelasticity. The examined electroviscoelastic systems are fitted with piezoelements whose electrodes are connected to the external RLC-circuits having resistance, capacitance and inductance. For electroviscoelastic systems as optimising parametres are in addition used: the properties of materials, location of the piezoelements, the conditions of their interaction, and the parameters of the external RLC-circuits. Optimization of the dynamic characteristics of the systems under study is realized in the context of the mechanical problem on natural vibrations.

1 Introduction

Investigation of the interaction between the fields of different physical nature is gaining great scientific and practical importance. One of the areas, where these studies have found wide use, is production and application of new composite materials known as smart materials.

Smart materials are widely applied for realization of different functional tasks. For example, structure shape control, detection of damage including the early-stage defects, control of dynamic processes, micropositioning, geometry control. There are many different industrial applications of smart composites on the basis of piezoelectric materials, which are very promising in the field of suppression of vibration and noise.

V.P. Matveenko (✉) · E.P. Kligman · M.A. Yurlov · N.A. Yurlova
Institute of Continuous Media Mechanics of the Ural Branch of RAS, Academician Korolev str., 1 Perm, 614013, Russia
e-mail: mvp@icmm.ru

H. Irschik et al. (eds.), *Advanced Dynamics and Model-Based Control of Structures and Machines*, DOI 10.1007/978-3-7091-0797-3_18,

In the framework of mathematical models, the main challenge in finding the optimal variants of application of smart composites on the basis of piezomaterials for suppressing (amplifying) the vibrations of the system is the development of highly effective methods of numerical analysis, which can be readily used to estimate of the dissipative properties of the examined structures. A qualitative estimation of the dissipative properties of the structure is generally based on the results of investigation of free and forced steady-state vibration.

Both these problems require time-consuming computations. In the case of free vibrations this analysis is based on the development of solutions to dynamic problems under different initial conditions, and in the forced vibration problems – on the development of solutions under different loading conditions for the prescribed frequency range of the disturbing influences.

As an alternative to the described approaches including those conventionally used for solving the optimization problems, we suggest a new formulation of the natural vibration problem, which allows us to estimate the damping properties of the system irrespective of external force, kinematic and other factors.

2 Mathematical formulation of the natural vibration problem

Let us consider the system as a piecewise homogeneous body of the volume $V = V_1 + V_2$, where the volume V_1 consists of homogeneous elastic and viscoelastic elements and the volume V_2 consists of piezoelectric elements. The piezoelectric elements can be connected through the electroded surface to the resistance, capacitance, and inductance elements.

The variational equation of motion of a body consisting of the viscous, viscoelastic and piezoelectric elements for an isothermal process can be written as [1–4]:

$$\begin{aligned}&\int\limits_{V_1}(\sigma_{ij}\delta\varepsilon_{ij} + \rho\ddot{u}_i\delta u_i)dV + \int\limits_{V_2}(\sigma_{ij}\delta\varepsilon_{ij} - D_i\delta E_i + \rho\ddot{u}_i\delta u_i)dV\\ &\quad - \int\limits_{\Omega_\sigma}\delta u_i P_i\, d\Omega - \int\limits_{\Omega_p} q_e\delta\phi d\Omega = 0\,,\end{aligned} \tag{1}$$

where σ_{ij}, ε_{ij} are the components of the stress and strain tensors, u_i are the components of the displacement vector, ρ is the specific density of the material, D_i, E_i are the components of the electric inductance vector and the vector of the electric field intensity; P_i are the components of the load vector; Ω_p is the surface bounding a piezoelectric element; q_e and ϕ are the surface charge density and the electric potential.

The electric field is considered to be potential, i.e. the condition $\phi_{,i} = -E_i$ is fulfilled.

For isothermal processes the elastic parts of the volume V_1 are determined by the Hooke law for isotropic or anisotropic material. The stresses and deformations in the viscoelastic parts of the composite body are related by the Boltzmann-Volterra relations

$$\sigma_{ij} = C^0_{ijkl}\varepsilon_{kl} - \int_0^t R_{ijkl}(t-\tau)\varepsilon_{kl}\, d\tau \tag{2}$$

For piezoelectric elements the following physical relations are valid:

$$\begin{aligned} \sigma_{ij} &= C_{ijkl}\varepsilon_{kl} - \beta_{ijk} E_k \\ D_k &= \beta_{ijk}\varepsilon_{ij} + e_{ki} E_i \end{aligned} \tag{3}$$

Here C^0_{ijkl} is the tensor of instant elastic constants, β_{ijk}, e_{kj} are the components of tensors of the piezoelectric and dielectric coefficients.

The deformations and displacements are related by the linear Cauchy equations.

For the problem of natural vibrations under homogeneous boundary conditions the solution is written in terms of displacements and stresses:

$$u_i(\mathbf{x}, t) = \bar{u}(\mathbf{x})\, e^{-i\omega t}\ , \tag{4}$$

where $\omega = \omega_R + i\omega_I$ is the complex natural vibration frequency. The real part of the complex natural frequency corresponds to the natural or resonance vibration frequency and the imaginary part - to the damping rate of the corresponding vibration mode.

For quasi-static processes described by relations (4) the physical relations (2) can be replaced by their complex analogue [5]:

$$\begin{aligned} \sigma_{ij} &\approx \left[C_{ijkl} - Q^c_{ijkl}(\omega_R) - iQ^s_{ijkl}(\omega_R)\right]\varepsilon_{kl} = \left[C^{(r)}_{ijkl}(\omega_R) + iC^{(i)}_{ijkl}(\omega_R)\right]\varepsilon_{kl} \\ &= C^*_{ijkl}(\omega_R)\varepsilon_{kl}\ , \end{aligned} \tag{5}$$

where C^*_{ijkl} is the tensor of complex dynamic moduli,

$$Q^c_{ijkl} = \int_0^\infty R_{ijkl}(\tau)\cos\omega_R\tau\, d\tau, \quad Q^S_{ijkl} = \int_0^\infty R_{ijkl}(\tau)\sin\omega_R\tau\, d\tau\ .$$

Taking into account of the homogeneous boundary conditions and solution (4), we can write the variational equation (1) as

$$\left[\int_V (\sigma_{ij}\delta\varepsilon_{ij} - \rho\omega^2\bar{u}_i\delta u_i)\, dV - \int_{V_2} D_i\delta E_i\, dV\right] e^{-i\omega t} = \int_{\Omega_p} q_e\delta\phi\, d\Omega \tag{6}$$

Let us consider the case when the electroded surface of the piezoelement is connected to the point of a zero potential by the element Z, which is a series circuit composed of the resistance R, capacitance C and inductance L elements

$$\phi = \frac{Q}{C} + RI + L\dot{I} = \frac{Q}{C} + R\dot{Q} + L\ddot{Q}\,, \quad Q = \int\limits_{\Omega_1} q\, d\Omega \tag{7}$$

where Q is the charge on the electrode, $I = \dot{Q}$ is the current in the conductor.

Since we examine a quasi-harmonic process, viz. $\phi(t) = \bar{\phi}e^{-i\omega t}$, the differential equations (7) can be solved with respect to Q:

$$Q(t) = \frac{\bar{\phi}e^{-i\omega t}}{C^{-1} - \omega^2 L - i\omega R}\,,$$

where $\omega = \omega_R + i\omega_I$ is the complex natural frequency of vibrations.

In view of equipotentiality of the electroded surface Ω_1 and the condition that the RLC-circuit is an external element, the following condition is satisfied:

$$\delta\phi \int\limits_{\Omega_1} q_e\, d\Omega = -\frac{\delta\phi\bar{\phi}_0 e^{-i\omega t}}{C^{-1} - \omega^2 L - i\omega R}\,,$$

where $\bar{\phi}_0$ is the potential on the electroded surface of the piezoelement Ω_1.

Finally, in the absence of other external forces equation of motion (6) can be written in the following form:

$$\int\limits_V (\sigma_{ij}\delta\varepsilon_{ij} - \rho\omega^2\bar{u}_i\delta u_i)\, dV - \int\limits_{V_2} D_i\delta E_i\, dV + \frac{\delta\phi\,\bar{\phi}_0}{C^{-1} - \omega^2 L - i\,\omega R} = 0 \tag{8}$$

Equation (8) is homogeneous and can be considered as a variational eigenvalue problem.

Nowadays, piezoelectrical elements are applied for constructing active smart-systems with the view to provide an effective suppression of vibrations and to extend the dynamic stability range of non-conservative systems. It is common practice to use for this purpose the electric potential feedbacks or its time derivatives. The existence of feedback of elements intended for digital processing of signals is the pro property that favors the construction of effective automatic control systems. Numerical simulation of active control systems in the framework of the continuum problem of electroviscoelasticity requires formulation of new types of the boundary conditions.

Let us rewrite (1) extending the surface integral.

$$\int_V (\sigma_{ij}\delta\varepsilon_{ij} - D_i\delta E_i)dV = \int_V (X_i - \rho\ddot{u}_i)\delta u_i\, dV + \int_{\Omega^0} (t_i\delta u_i + \delta\phi\, q_e)d\Omega + \int_{\Omega^{el}} t_i\delta u_i\, d\Omega + \sum_{(k\neq i,j)} \delta\phi_k Q_k \tag{9}$$

Here Ω^0 is the non-electroded surface, $\Omega^{el} = \sum_{k=1}^{n} \Omega_k^{el}$ is the electroded surface, consisting of n parts, ϕ_k and $Q_k = \int_{\Omega_k^{el}} q_e\, d\Omega$ are the potential and the charge on the k-th electrode. Keeping within the generality of the approach we consider the case when the charge $Q_i = \alpha\phi_j$, (where α is the feedback coefficient) is supplied to the electrode i. Then, (9) takes the following form:

$$\int_V (\sigma_{ij}\delta\varepsilon_{ij} - D_i\delta E_i)\, dV = \int_V (X_i - \rho\ddot{u}_i)\delta u_i\, dV + \int_{\Omega^0} (t_i\delta u_i + \delta\phi q_e)\, d\Omega + \int_{\Omega^{el}} t_i\delta u_i\, d\Omega + \sum_{(k\neq i,j)} \delta\phi_k Q_k + \delta\phi_i\alpha\phi_j + \delta\phi_j Q_j \tag{10}$$

The capabilities of the advanced electronic facilities allow us to "measure" the electric potential on the electrode leaving its charge practically intact, viz. $Q_j = 0$.

$$\int_V (\sigma_{ij}\delta\varepsilon_{ij} - D_i\delta E_i)\, dV = \int_V (X_i - \rho\ddot{u}_i)\delta u_i\, dV + \int_{\Omega^0} (t_i\delta u_i + \delta\phi q_e)\, d\Omega + \int_{\Omega^{el}} t_i\delta u_i\, d\Omega + \sum_{(k\neq i,j)} \delta\phi_k Q_k + \delta\phi_i\alpha\phi_j \tag{11}$$

If the body is free of the external mechanical and electrical loads, the variational equation (10) becomes homogeneous and can be interpreted as an eigenvalue problem.

$$\int_V (\sigma_{ij}\delta\varepsilon_{ij} - D_i\delta E_i)\, dV + \int_V \rho\ddot{u}_i\delta u_i\, dV - \delta\phi_i\alpha\phi_j = 0 \tag{12}$$

To sum up it should be noted that the solution of the natural vibration problem for electroviscoelastic bodies can be used to obtain the required number of complex natural vibration frequencies.

Here, as optimization parameters we can use the geometry of elastic, viscoelastic and piezoelectric elements of the composite body, the boundary conditions, characteristics of the external electric circuit, feedback coefficients, etc.

3 Results of calculations

Numerical realization of the variational eigenvalue problem is based on the algorithm, in which the desired solution is represented as a linear combination of the natural vibration modes of the corresponding elastic body without external circuits which are found by using the ANSYS package. Let us consider a trapezoidal plate (Fig. 1), simulating an airplane wing, as one of the possible application of the natural vibration problem for viscoelastic bodies with the external electric circuits, which we used to estimate the dissipative properties of the structure.

Computations were made for the following dimensions: $l_1 = 9,000$ mm, $l_2 = 15,000$ mm, $l_3 = 3,000$ mm, $l_4 = 5,700$ mm, $l_5 = 4,050$ mm. The plate thickness and the thickness of a piezomaterial layer are equal to 300 mm and 150 mm, respectively. The plate was made of metal with the mechanical characteristics $E = 2 \cdot 10^{11}$ Pa, $\nu = 0.3$. For piezomaterial we chose piezoceramic PZT-4 showing the standard properties [6]. The results of computation were represented in terms of the vibration characteristics: $f = \omega_R/2\pi$ is the resonance frequency (HZ), $N = \omega_I$ is the damping coefficient (c^{-1}).

Figure 2 shows the results of numerical simulation for the first vibration mode in the case when the external electric circuit includes only one RLC element

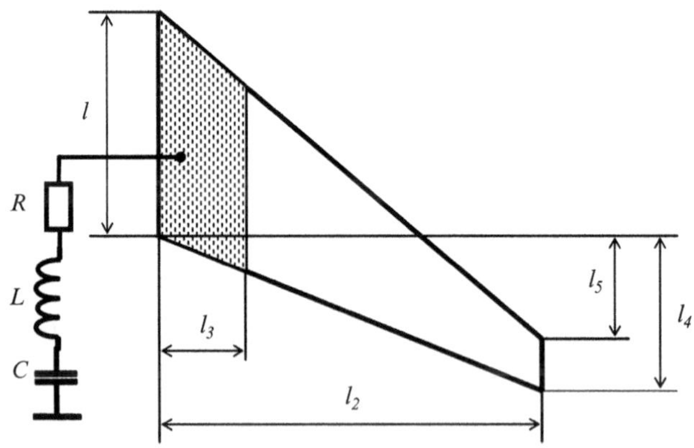

Fig. 1 Computational scheme of the trapezoidal plate with built-in piezoelectric element and the external electric circuit

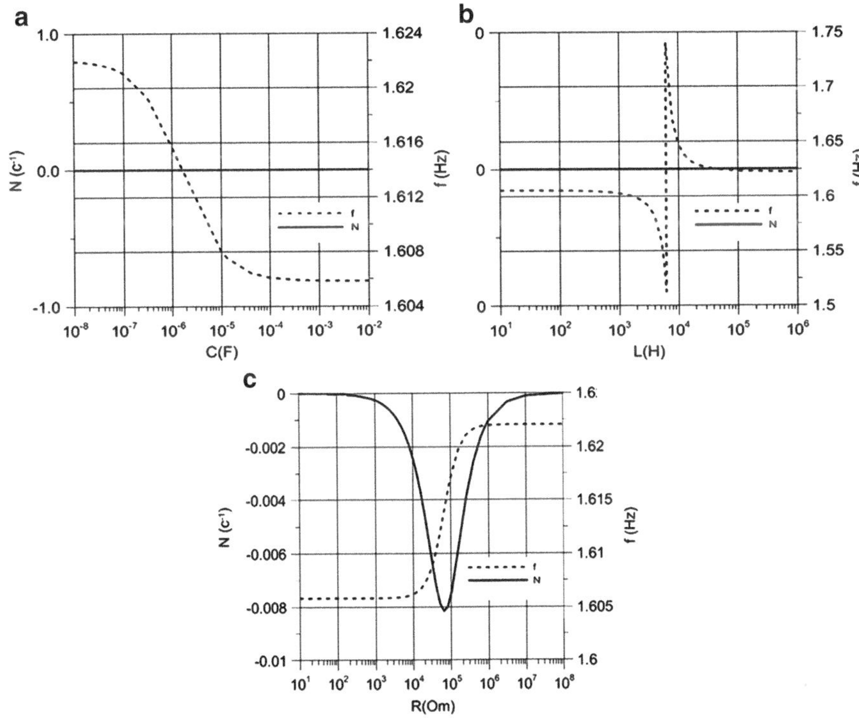

Fig. 2 Resonance frequency and damping coefficient as a function of capacitance (**a**), inductance (**b**) and resistance (**c**)

(the capacity, inductance or resistance element) for different dimensions of the element used. The results show that the damping coefficient in simple electric circuits is affected only by passive resistance.

The results of numerical simulation for the first vibration mode are given in Fig. 3. Computations were made for the circuit consisting of series connected resistance and inductance elements. Here the damping coefficient is plotted versus resistance for different values of inductance. It is seen that maximum damping is provided at $L = 6,340$ H.

A comparison of the damping coefficients calculated for a simple R-circuit and series connected RL-circuit clearly demonstrates that the latter is 15 times more effective. This can be explained by the fact that capacitance of the piezoelement and inductance form a resonant circuit, which essentially increases the electric current flowing though the circuit.

The analysis of the analogous relationships for the second, third and forth modes confirms their qualitative identity with the first mode.

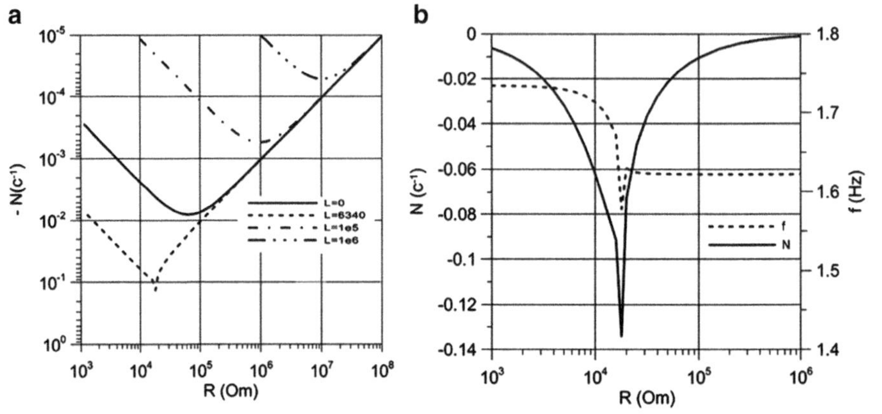

Fig. 3 Damping coefficient vs. resistance for a system with external series-connected RL-circuit at different values of inductance (**a**) series-connected RL-circuit, $L_{opt} = 6{,}340$ H; (**b**) parallel RL-circuit, $L_{opt} = 6{,}500$ H

4 Conclusion

The approach developed in this paper is based on the modal analysis of electroviscoelastic systems with external passive and active discrete couplings. The numerical simulations made with this approach support its validity as a universal and effective means for determination and optimization of dynamic characteristics of structures made of smart materials.

A new mechanical problem on natural vibrations of viscoelastic bodies with the external circuits has been solved by applying the ANSYS software package which has demonstrated its great potentialities for numerical simulation of objects of different spatial configuration.

Acknowledgements This study was supported by the RFBR grant (project 09-01-92651) and the program of the Presidium of RAS 09-P-1-1010.

References

1. Nashif, A., Jones, D., Henderson, J.: Vibration Damping. MIR, Moscow (1988)
2. Vasidzu, K.: Variational Methods in Elasticity and Plasticity Theory. MIR, Moscow (1987)
3. Parton, V.Z., Kudryavtsev, B.A.: Electromagnetoelasticity of Piezoelectric and Electrically Conducting Bodies. Nauka, Moscow (1988)
4. Karnaukhov, V.G., Kirichok, I.F.: Electrothermoviscoelasticity. Nauk.Dumka, Kiev (1988)
5. Adamov, A.A., Matveenko, V.P., Trufanov, N.A., Shardakov, I.N.: Methods of applied viscoelasticity. UB RAS, Ekaterinburg (2003)
6. Inman, D.J.: Smart structures solutions to vibration problems. Proc. ISMA 23: Leuven, Belgium, **1**, 1–12 (1998)

Control of Surface Waviness

N.F. Morozov and P.E. Tovstik

Abstract The elastic material loses the bulk stability at sufficiently large compression when the Hadamard condition is not satisfied. An elastic half-space could lose at compression the surface stability with the buckling waves localized near a free plane and exponentially decreased away from it. The compressed elastic plate lying on an elastic foundation also could lose stability. The problems of the critical loads and buckling modes determination are discussed.

1 Introduction

Problems of stability of boundaries of the elastic domains and interfaces are important for different applications. In this paper we investigate the stability of an elastic half-space and a plate on an elastic foundation under compression loads.

It is known that the elastic material loses the bulk stability at sufficiently large compression when the Hadamard condition is not satisfied [1–3]. The loss of stability of elastic half-space could be accompanied by the formation of buckling waves localized near a free plane and exponentially decreased away from it [2–4]. The compressed elastic plate lying on an elastic foundation could lose stability at compression [5–7]. For these problems the critical compression and the buckling modes for a transversely isotropic linearly elastic material are established. The critical compression is to be found from the equations of bifurcation. But the corresponding buckling modes could not be determined uniquely.

Let us consider a compressed transversely isotropic material in the space $Ox_1x_2x_3(x_3 = z)$ in which the horizontal planes $z =$ const are the planes of

N.F. Morozov (✉) · P.E. Tovstik
Saint-Petersburg State University, Universitetskaya nab.7/9, Saint-Petersburg, Russia
e-mail: morozov@nm1016.spb.edu; Peter.Tovstik@mail.ru

H. Irschik et al. (eds.), *Advanced Dynamics and Model-Based Control of Structures and Machines*, DOI 10.1007/978-3-7091-0797-3_19,

isotropy. We assume that the constant initial horizontal deformations $\partial u_1^0/\partial x_1 = \partial u_2^0/\partial x_2 = -e$ are given, and the initial vertical stresses are equal to zero ($\sigma_{33}^0 = 0$). These deformations generate the initial horizontal stresses $\sigma_{11}^0 = \sigma_{22}^0 = \sigma_0$ and the initial vertical deformation $\partial u_3^0/\partial z = e_3 > 0$.

It is assumed that the elastic modules could depend only on the vertical coordinate z, and we look for the solution of the bifurcation equations in the form

$$u_k(x_1, x_2, x_3) = u_k(z)\exp\{i(r_1x_1 + r_2x_2)\}, \quad k = 1, 2, 3, \quad i = \sqrt{-1}\,, \tag{1}$$

where u_k are the additional displacements at bifurcation and r_j are unknown wave numbers.

The bifurcation equations contain only $r = \sqrt{r_1^2 + r_2^2}$ but r_1 and r_2 are not contained in these equations separately. The bifurcation system of the 6th order may be divided into one equation of the 2nd order and into the system of the 4th order. Two kinds of buckling modes for volume stability loss are possible. The kind 1 is the rotation mode with the displacements in the plane of isotropy ($u_1, u_2 \neq 0$, $u_3 = 0$), and the critical compression $e^{(1)}$ may be found from the equation of the 2nd order. For the kind 2 the displacements in all the directions are not equal to zero ($u_1, u_2, u_3 \neq 0$), and the critical compression $e^{(2)}$ may be found from the system of the 4th order. Which of the critical values, $e^{(1)}$ or $e^{(2)}$, is smaller (or which of the buckling modes takes place), depends on the elastic modules of the material. If the elastic properties of the material are close to the isotropic ones then the kind 1 is implemented, and, on contrary, if the material is soft in the Oz direction then the kind 2 takes place. If the material is a homogeneous one (the elastic modules are constant) then $u_k(z) = u_k^0\, e^{ir_3z}$, and the buckling mode occupies the entire space. In this case the critical compression does not depend on the wave number r.

For heterogeneous material (the elastic modules depend on z) the buckling mode is localized near the weakest plane $z = z_*$, and the critical compression depends on the wave number $r (e^{(k)} = f(r))$. If the modules are continuous functions in z then the function $f(r)$ decreases and $f(r) \to f_\infty$ at $r \to \infty$. f_∞ is equal to the critical compression of a homogeneous material with the elastic modules coinciding with the values of a heterogeneous material at $z = z_*$.

2 Pre-buckling state and its bifurcation

The elastic potential energy density for an orthotropic material is

$$U = \frac{1}{2}\sum_{i,j=1}^{3} E_{ij}\varepsilon_{ii}\varepsilon_{jj} + \sum_{i\neq j} G_{ij}\varepsilon_{ij}^2, \tag{2}$$

where Cauchy–Green deformations ε_{ij} are equal

$$\varepsilon_{ij} = \frac{1}{2}\left(\frac{\partial u_i}{\partial x_j} + \frac{\partial u_j}{\partial x_i} + \sum_{k=1}^{3}\frac{\partial u_k}{\partial x_i}\frac{\partial u_k}{\partial x_j}\right), \tag{3}$$

and u_i are the projections of displacements, E_{ij}, G_{ij} are the elastic modules. For a transversely isotropic material the following restrictions are valid

$$E_{11} = E_{22} = E_{12} + 2G_{12}, \quad E_{13} = E_{23}, \quad G_{13} = G_{23}. \tag{4}$$

We study the homogeneous plane pre-buckling state in the non-linear approach described by the relations

$$e_i = \frac{\partial u_i^0}{\partial x_i}, \quad \varepsilon_{ii}^0 = e_i + \frac{1}{2}e_i^2, \quad \sigma_i^0 = \sum_{k=1}^{3} E_{ik}\varepsilon_{kk}^0, \quad i = 1,2,3, \tag{5}$$

where u_i^0 are the pre-buckling displacements, and assume that

$$e_1 = e_2 = -e, \qquad \sigma_3^0 = 0, \qquad \varepsilon_{11}^0 = \varepsilon_{22}^0 = -e + \frac{e^2}{2}, \qquad \varepsilon_{ij}^0 = 0, \quad i \neq j,$$

$$\sigma_1^0 = \sigma_2^0 = \sigma^0 = \left(E_{11} + E_{12} - \frac{2E_{13}^2}{E_{33}}\right)\varepsilon_{11}^0, \quad e_3 = \sqrt{1 - \frac{4E_{13}}{E_{33}}\varepsilon_{11}^0} - 1. \tag{6}$$

We introduce small perturbations v_i of the pre-buckling state by relations

$$u_i = u_i^0 + l_i v_i\,, \quad l_i = 1 + e_i. \tag{7}$$

Then we get deformations

$$\varepsilon_{ij} = \varepsilon_{ij}^0 + \frac{1}{2}\left(l_i^2\frac{\partial v_i}{\partial x_j} + l_j^2\frac{\partial v_j}{\partial x_i} + \frac{1}{2}\sum_{k=1}^{3} l_k^2\frac{\partial v_k}{\partial x_i}\frac{\partial v_k}{\partial x_j}\right) \tag{8}$$

and expand the potential (2) by the powers of v_i

$$U = U_0 + U_1 + U_2 + O(v_i^3), \quad U_1 = A_{ik}\frac{\partial v_i}{\partial x_k}, \quad U_2 = A_{ikjn}\frac{\partial v_i}{\partial x_k}\frac{\partial v_j}{\partial x_n}, \tag{9}$$

where A_{ik} and A_{ikjn} are the constant coefficients (the summation in the repeated indexes is supposed). The variation of the functional U_2 after integration by parts leads to the equations of equilibrium bifurcation

$$M_k(\mathbf{v}) = \left(E_{kk}l_k^2 + \sigma_k^0\right)\frac{\partial^2 v_k}{\partial x_k^2} + \sum_{i\neq k}\left((G_{ki}l_k^2 + \sigma_i^0)\frac{\partial^2 v_k}{\partial x_i^2}\right.$$

$$\left. + \ (E_{ki} + G_{ki})l_i^2 \frac{\partial^2 v_i}{\partial x_k \partial x_i}\right) = 0, \quad k = 1, 2, 3. \tag{10}$$

In the non-linear approach the multipliers l_j take the pre-buckling deformations into account. In the linear approach we put $l_j = 1$ and system (10) coincides with the system investigated in [2–4] where the pre-buckling strains and stresses are found in a linear approximation.

We look for solutions of system (10) as a double periodic functions of the horizontal co-ordinates x_1, x_2

$$\begin{aligned} v_1(x_1, x_2, z) &= v_1(z)\cos(r_1 x_1)\sin(r_2 x_2), \\ v_2(x_1, x_2, z) &= v_2(z)\sin(r_1 x_1)\cos(r_2 x_2), \\ w(x_1, x_2, z) &= w(z)\sin(r_1 x_1)\sin(r_2 x_2), \qquad w = v_3. \end{aligned} \tag{11}$$

We introduce the auxiliary unknown functions to simplify system (10) [5].

$$U = (r_1 v_1 + r_2 v_2)/r, \quad V = (r_2 v_1 - r_1 v_2)/r, \qquad r^2 = r_1^2 + r_2^2. \tag{12}$$

Then system (10) divides into two parts

$$\left(G_{13}l^2 V'\right)' - r^2\left(G_{12}l^2 + \sigma^0\right)V = 0, \qquad l_1 = l_2 = l, \quad ()' = \frac{d()}{dz}; \tag{13}$$

$$\begin{aligned} &\left(G_{13}l^2 U'\right)' - (E_{11}l^2 + \sigma^0)r^2 U + (G_{13}l_3 r w)' + E_{13}l_3 r w' = 0, \\ &-\left(rE_{13}l^2 U\right)' - G_{13}l^2 U' + \left(E_{33}l_3^2 w'\right)' - r^2\left(G_{13}l_3^2 + \sigma^0\right)w = 0. \end{aligned} \tag{14}$$

Equation (13) describes the rotation deformations with displacements in the plane of isotropy Ox_1x_2, and system (14) describes deformations with displacements in three directions. We refer to the corresponding buckling modes as to the modes of the kind 1 and of the kind 2.

3 Surface stability

We study the stability of a transversely isotropic homogeneous half-space $x_3 \le 0$ under compression e. The same equations (13) and (14) may be used. At the free plane $x_3 = z = 0$ the boundary conditions $\sigma_{k3} = 0$ are fulfilled or

$$\frac{\partial V}{\partial z} = 0, \tag{15}$$

$$l^2\frac{\partial U}{\partial z} + l_3^2\Delta w = 0, \qquad E_{13}l^2U + E_{33}l_3^2\frac{\partial w}{\partial z} = 0. \tag{16}$$

At $z \to -\infty$ the decreasing conditions are fulfilled $\{V, U, w\} \to 0$. Equation (13) does not have any non-zero bounded solutions satisfying condition (15) and decreasing condition at $z \to -\infty$.

We look for solutions of system (14) in the form

$$U(z) = U_0 e^{r\alpha z}, \quad w(z) = w_0 e^{r\alpha z}, \qquad Re(\alpha) > 0. \tag{17}$$

Then system (14) gives

$$\begin{aligned} &\left(G_{13}l^2\alpha^2 - E_{11}l^2 - \sigma^0\right)U_0 - (E_{13} + G_{13})l_3^2\alpha w_0 = 0,\\ &(E_{13} + G_{13})l^2\alpha U_0 + \left(E_{33}l_3^2\alpha^2 - G_{13}l_3^2 - \sigma^0\right)w_0 = 0, \end{aligned} \tag{18}$$

and the boundary conditions (16) may be rewritten in the form

$$l^2\alpha U_0 - l_3^2 w_0 = 0, \qquad E_{13}l^2U_0 + E_{33}l_3^2\alpha w_0 = 0. \tag{19}$$

The parameter α satisfies equation

$$\left(G_{13}\alpha^2 - E_{11} - \sigma^0 l^{-2}\right)\left(E_{33}\alpha^2 - G_{13} - \sigma^0 l_3^{-2}\right) + (E_{13} + G_{13})^2\alpha^2 = 0. \tag{20}$$

If $G_{13}l_3^2 + \sigma^0 > 0$ (20) has two roots α_1 and α_2 with $Re(\alpha_k) > 0$ and solution of system (14) satisfying the decreasing conditions at $z \to -\infty$ is

$$w = \sum_{k=1,2} C_k e^{r\alpha_k z},\; U = \sum_{k=1,2} C_k b_k e^{r\alpha_k z},\; b_k = \frac{G_{13}l_3^2 + \sigma^0 - E_{33}l_3^2\alpha_k^2}{(E_{13} + G_{13})l^2\alpha_k}. \tag{21}$$

The arbitrary constants C_k may be found from boundary conditions (19). As a result we get the equation for critical stress σ^0 or for the initial critical compression $e^{(s)}$

$$\frac{E_{13}l^2b_1 + E_{33}l_3^2\alpha_1}{l^2\alpha_1 b_1 - l_3^2} = \frac{E_{13}l^2b_2 + E_{33}l_3^2\alpha_2}{l^2\alpha_2 b_2 - l_3^2}. \tag{22}$$

We can see that (22) does not contain the wave numbers r_1, r_2 therefore as for the volume stability loss the buckling mode is indefinite. The critical compressions $e^{(s)}$ for the surface loss of stability as well as for the volume loss of stability $e^{(1)}$ and $e^{(2)}$ (in order to make a comparison) are compared. According to the general variation principle (in the presence of the additional restrictions the eigenvalues are larger) the following inequality

$$e^{(s)} < \min\left\{e^{(1)}, e^{(2)}\right\} \tag{23}$$

is valid.

4 Stability of a compressed plate on an elastic foundation

On the contrary to the previously discussed problems, for this problem could be found all the parameters of the buckling modes. An approximate solution would be constructed.

We study stability of an infinite compressed plate of thickness h lying on the elastic foundation. For simplicity we assume that as the plate so the foundation are made of an isotropic material with the Young modules $E,\ E_0$ and the Poisson ratios $\nu,\ \nu_0$ where index 0 is used for the foundation. The homogeneous initial horizontal compression e of a plate is given.

To describe the plate additional deformations the Kirchhoff–Love approximation is used. Then the normal deflection $w(x_1, x_2)$ of the middle plane satisfies to equation [4, 6]

$$D\Delta\Delta w - T\Delta w + \sigma_{33}^0 = 0, \qquad \Delta w = \frac{\partial^2 w}{\partial x_1^2} + \frac{\partial^2 w}{\partial x_2^2}. \tag{24}$$

$$T = h\sigma_0 = -\frac{Ehe}{1-\nu}, \qquad D = \frac{Eh^3}{12(1-\nu^2)}, \tag{25}$$

and σ_{33}^0 is the stress between the plate and the foundation.

We look for the solution of (24) in the form

$$w(x_1, x_2) = w_0 \cos(r_1 x_1) \cos(r_2 x_2) \tag{26}$$

with unknown wave numbers r_1 and r_2. Then it is easy to find the stress σ_{33}^0 in assumption that between the plate and the foundation there is a full contact, and that the foundation has an infinite depth [9, 10].

$$\sigma_{33}^0 = c_{33} r w_0, \qquad c_{33} = \frac{E_0(1-\nu_0)}{(1+\nu_0)(3-4\nu_0)}, \qquad r = \sqrt{r_1^2 + r_2^2}. \tag{27}$$

Then from (24) we find

$$e = \frac{(1-\nu)(Dr^3 + c_{33})}{Ehr}, \qquad -\sigma_0 = F_1(r) = \frac{Dr^3 + c_{33}}{hr}. \tag{28}$$

By minimizing e in r we find critical values e_* and r_*.

$$(r_* h)^3 = \frac{6(1-\nu^2)(1-\nu_0)}{(1+\nu_0)(3-4\nu_0)} \frac{E_0}{E}, \qquad -e_* = \frac{(1-\nu)(1-\nu_0)}{(1+\nu_0)(3-4\nu_0)} \frac{E_0}{E r_* h}. \tag{29}$$

The typical length of buckling waves L is equal to $L = 2\pi/r$ and Kirchhoff–Love hypotheses are acceptable only if $h \ll L$ or $rh \ll 1$. To fulfill this inequality it is necessary to suppose that the foundation is soft enough ($E_0 \ll E$). Therefore the length of waves L is found but from the bifurcation equation (24) the buckling mode depending on the relation between the wave numbers r_1 and r_2 remains indefinite. The comparison of approximate relations (30) with exact numerical solution shows that exactness of the relations (29) grows when the ratio E_0/E decreases.

5 Post-critical behavior of a plate on an elastic foundation

To find the relation between r_1 and r_2 we study approximately the initial post-critical behavior of a plate. We calculate the elastic energy density using non-linear expressions for strains in the plate with Kirchhoff–Love hypotheses and linear expressions for strains in the foundation. The elastic energy density of plate U_p is equal to

$$U_p = \frac{E}{2L_1L_2}\int_0^{L_1} dx_1 \int_0^{L_2} dx_2 \int_{-h/2}^{h/2} \left(\varepsilon_{11}^2 + 2\nu\varepsilon_{11}\varepsilon_{22} + \varepsilon_{22}^2 + 2(1-\nu)\varepsilon_{12}^2\right) dz, \tag{30}$$

$$\text{where } \varepsilon_{ij} = \frac{1}{2}\left(\frac{\partial u_i}{\partial x_j} + \frac{\partial u_j}{\partial x_i} - 2z\frac{\partial^2 w}{\partial x_i \partial x_j} + \frac{\partial w}{\partial x_i}\frac{\partial w}{\partial x_j}\right). \tag{31}$$

We take $w(x_1, x_2)$ in form (26) and choose the horizontal displacements $u_i(x_1, x_2)$ of the plate middle plane so that the energy U_p be minimal. Then we get [4]

$$u_i(x_1, x_2) = \frac{r_i w_0^2}{16}\sin 2r_i x_i\,(1+\cos 2r_j x_j), \quad i, j = 1, 2, \quad i \neq j. \tag{32}$$

The energy density of foundation U_f is equal to

$$U_f = \frac{c_{33} r w_0^2}{8}. \tag{33}$$

The total energy density $U = U_p + U_f$ may be written in the form [4]

$$U = \frac{r^2 h w_0^2}{8}(F_1(r) + \sigma_0) + F_2(r_1, r_2) w_0^4, \;\; F_2 = \frac{Eh(3r_1^4 + 4\nu r_1^2 r_2^2 + 3r_2^4)}{256(1-\nu^2)}, \tag{34}$$

where $F_1(r)$ is the same as in (8) and the critical initial compression σ_* is equal to $\sigma_* = -F(r_*)$. If $|\sigma_0| < |\sigma_*|$ then the energy U is minimal at $w_0 = 0$. If $|\sigma_0| > |\sigma_*|$ then this energy is minimal at

$$w_0^2 = \frac{(\sigma_* - \sigma_0) r_*^2 h}{16F_2}, \quad r_1 = r_2 = \frac{r_*}{\sqrt{2}}, \quad F_2 = \frac{(3+2\nu)E r_*^4}{64(1-\nu^2)}. \tag{35}$$

Therefore in this problem the buckling mode is a chessboard-like one (see also [8]).

6 Conclusion

The surface stability of a compressed transversely isotropic homogeneous half-space $z \leq 0$ with the free plane $z = 0$ and stability of plate on elastic foundation are considered. For the surface stability loss only the kind 2 of the buckling mode which is described by the system of the 4th order is possible. As in the case of the volume stability the bifurcation equation contains only $r = \sqrt{r_1^2 + r_2^2}$ but the critical compression $e^{(s)}$ does not depend on r. The free surface $u_3 = u_3(x_1, x_2, 0)$ after buckling is a double periodic function of x_1 and x_2. But as the wave length $L = 2\pi/r$ so the buckling mode which is described by the relation between r_1 and r_2 remains indefinite.

The buckling problem of the infinite compressed plate lying on the soft elastic foundation has a much more definite solution. In a linear approximation the buckling mode which is described by the relation between r_1 and r_2 is still indefinite. In [4, 7] the post-buckling behavior of the compressed plate on the elastic foundation is investigated and it is shown that the minimal elastic energy density corresponds to the case with $r_1 = r_2$. In all these problems, especially in the cases without a plate, the initial critical strains are large enough.

Acknowledgements The paper is written with the financial support of Russian Foundation of Basic Investigations, grants 09.01.92002, 09.01.00642 and 10.01.00244.

References

1. Ciarlet, P.S.: Mathematical Elasticity. North-Holland, Amsterdam etc (1988)
2. Morozov, N.F., Tovstik, P.E.: On the modes of the surface stability loss. Problems of nonlinear mechanics of a deformable rigid body, conference 2, pp. 270–273, Kazan (2009)
3. Tovstik, P.E.: The volume and the surface stability of a transversely isotropic material under compression. Vestnik St.Petersburg University Series 1, N 1 (2010)
4. Morozov, N.F., Paukshto, M.V., Tovstik, P.E.: Stability of the surface layer under thermal loading. Izv. Russ. Akad. Nauk. MTT. (1), 130–139 (1998)
5. Tovstik, P.E.: Local stability of plates and shallow shells on an elastic foundation. Izv. Russ. Akad. Nauk. MTT. (1), 147–160 (2005)
6. Tovstik, P.E.: Vibrations and stability of a pre-stressed plate on an elastic foundation. Prikl. Math. Mekh. **73**(1), 106–120 (2009)
7. Morozov, N.F., Tovstik, P.E.: On the buckling modes of plate on an elastic foundation. Izv. Russ. Akad. Nauk. MTT. (4), 30–42 (2010)
8. Panin, L.E., Panin, V.E.: Effect of the "chessboard" and mass transfer in interfacial media of organic and inorganic nature. Phys. Nanomech. **10**(6), 5–20 (2007)
9. Tovstik, P.E., Smirnov, A.L.: Asymptotic Methods in the Buckling Theory of Elastic Shells, p. 347. World Scientific, Singapore, New Jersey, London, Hong Kong (2007)
10. Il'gamov, M.A., Ivanov, V.A., Gulin, B.V.: Strength, Stability and Dynamics of Shells with an Elastic Filler. Nauka, Moscow (1977)

Nonlinear Finite Element Modelling of Moving Beam Vibrations Controlled by Distributed Actuators

K. Nachbagauer, C. Zehetner, and J. Gerstmayr

Abstract In many applications, nonlinear beams undergoing bending, axial and shear deformation are important structural elements. In the present paper, a shear deformable beam finite element is presented for such applications. Since displacements and displacement gradients are chosen as the nodal degrees of freedom, an equivalent displacement and rotation interpolation is retrieved. The definition of strain energy is based on Reissner's nonlinear rod theory with special strain measures for axial strain, shear strain and bending strain. Furthermore, a thickness deformation is introduced by adding an according term to the virtual work of internal forces. This underlying formulation is extended for piezo-electric actuation. The obtained beam finite elements are applied to a two-link robot with two flexible arms with tip masses. Distributed and concentrated masses cause flexural vibrations, which are compensated by means of piezo-electric actuators attached to the arms. A numerical example of a highly flexible robot with piezo-electric actuation and feedforward control is presented to show the applicability of the finite element.

1 Introduction

For the modeling of large deformation structural problems in two and three space dimensions, two mainly used theories should be mentioned here: Reissner's nonlinear rod theory [9] and the absolute nodal coordinate formulation (ANCF) by Shabana [10], [11]. The main conceptual difference in the ANCF elements is

K. Nachbagauer (✉)
Institute of Technical Mechanics, Johannes Kepler University Linz, Altenbergerstraße 69, 4040 Linz, Austria
e-mail: karin.nachbagauer@jku.at

C. Zehetner · J. Gerstmayr
Linz Center of Mechatronics GmbH, Altenberger Straße 69, 4040 Linz, Austria
e-mail: christian.zehetner@lcm.at; johannes.gerstmayr@lcm.at

H. Irschik et al. (eds.), *Advanced Dynamics and Model-Based Control of Structures and Machines*, DOI 10.1007/978-3-7091-0797-3_20,

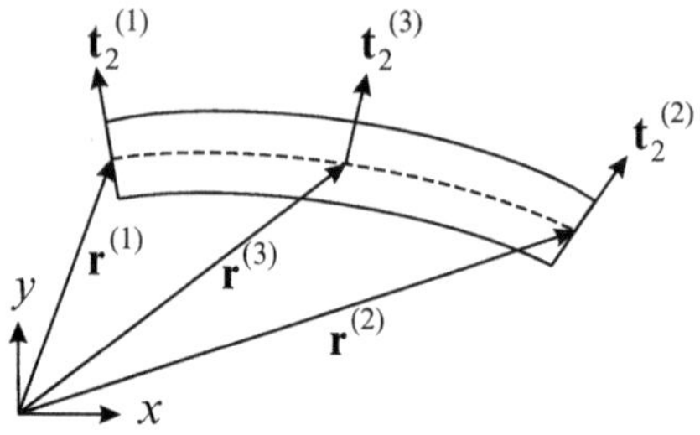

Fig. 1 Position vectors and slopes of the quadratic planar shear deformable ANCF beam element

the absence of rotational degrees of freedom, while it can be shown [5] that the behaviour of the obtained ANCF element is fully equivalent to that of the original element by Simo and Vu-Quoc [12]. In the present paper, the ANCF and the classical nonlinear rod theory are unified in order to describe a quadratic shear deformable beam element, see Fig. 1. The deformation energy is computed according to Simo and Vu-Quoc [12]. As an application of this quadratic ANCF finite element, a flexible multibody system is investigated. The main subject is the control of vibrations using piezo-electric patches as sensors and actuators. The advantage of the absolute nodal coordinate formulation is the availability of the interpolation of the displacement field, and the variety of possibilities to model the work of internal or external forces in a straight forward manner [2]. The contribution of a piezo-electric actuation by means of patches is given w.r.t. the virtual work of internal forces, see "Work of Elastic Forces and Piezo-electric Actuation". In "Numerical Example: A Highly Flexible Two-link Robot" a two-link robot consisting of two flexible arms with concentrated tip masses is considered, in which flexible vibrations are caused by inertial forces. For an alternative modelling and further information on the control schemes, see [13].

2 Geometric description and definition of the finite element

In the present section, the geometric setup of a shear deformable beam finite element is defined. The shear deformation is taken into account by means of a directional vector to model the deformation of the cross section. More precisely, the transverse gradient vector or slope vector, which points in direction of the cross section, is used. The position vector of the element nodes represents the position of the beam axis, while the direction of the transverse gradient models shear deformation, the length of the slope vector represents thickness deformation. Four different configurations of the element are considered, as shown in Fig. 2. The transformation from the unit element (a) to the scaled straight element (b) leads to the scaling factors $\xi = L\xi_0/2$ and $\eta = H\eta_0/2$. The position vector of the local point (ξ, η) in the undeformed geometry is denoted by $\mathbf{r}_0(\xi, \eta)$, which corresponds to the position vector $\mathbf{r}(\xi, \eta)$ in the deformed geometry, see Fig. 2. The displacement vector $\mathbf{u}(\xi, \eta)$ follows the relation

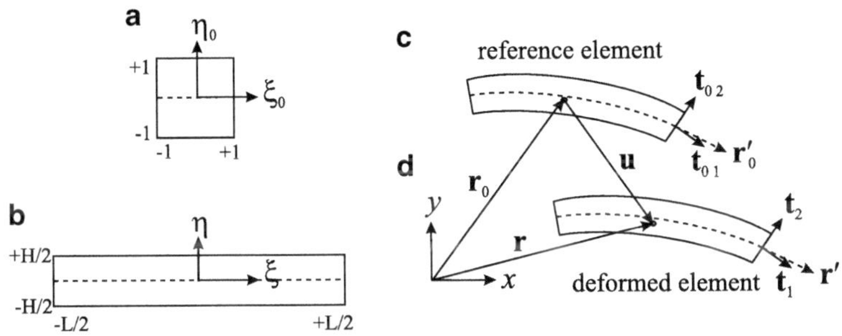

Fig. 2 (**a**) Unit element, (**b**) straight scaled reference element in unit configuration, in which the shape functions will be defined and which allows the definition of slope vectors, (**c**) reference element and (**d**) deformed element in deformed configuration

$$\mathbf{r}(\xi, \eta) = \mathbf{r}_0(\xi, \eta) + \mathbf{u}(\xi, \eta). \tag{1}$$

In the following, primes stand for the derivative with respect to the local axial coordinate ξ, $\mathbf{r}' = \frac{\partial \mathbf{r}}{\partial \xi}$. The slope vectors of the beam cross section are defined by the displacement gradients with respect to the transverse coordinate η, which are denoted by $\mathbf{r}_{,\eta} = \frac{\partial \mathbf{r}}{\partial \eta}$. The position vector of the axis $\mathbf{r}$ is written as $\mathbf{r}(\xi) = \mathbf{r}(\xi, 0)$. The tangent direction of the beam centerline is defined by $\mathbf{r}'$. The vector $\mathbf{t}_2$ points in direction of the deformed beam cross section, while $\mathbf{t}_1$ is perpendicular to $\mathbf{t}_2$. All quantities corresponding to the reference configuration are indexed with subscript zero, e. g. $\mathbf{t}_{0\,1}$ in the reference configuration belongs to $\mathbf{t}_1$ in deformed configuration.

In the following, a quadratic, three-noded ANCF element is defined, see Fig. 1. An analogous two-noded linear ANCF element can be found in [5]. In contrast to standard ANCF elements, in which position and slope vectors are chosen, the respective displacement-based quantities, i. e. the nodal displacements and change of slope vectors are used here. Four element coordinates are defined in each node n, namely the components of the nodal displacement $\mathbf{u}^{(n)}$ and the change of the slope vector $\mathbf{u}_{,\eta}^{(n)}$. The displacements $\mathbf{u}^{(1)}, \mathbf{u}^{(2)}$ and $\mathbf{u}^{(3)}$ and displacement gradients $\mathbf{u}_{,\eta}^{(1)}, \mathbf{u}_{,\eta}^{(2)}$ and $\mathbf{u}_{,\eta}^{(3)}$ at the nodal points are chosen as degrees of freedom, leading to the vector of unknowns

$$\mathbf{q} = [\mathbf{u}^{(1)\,T} \quad \mathbf{u}_{,\eta}^{(1)\,T} \quad \mathbf{u}^{(2)\,T} \quad \mathbf{u}_{,\eta}^{(2)\,T} \quad \mathbf{u}^{(3)\,T} \quad \mathbf{u}_{,\eta}^{(3)\,T}]^T. \tag{2}$$

The reference element is described by the generalized coordinate vector $\mathbf{q}_0$. The displacement vector $\mathbf{u}$ is interpolated quadratically by six shape functions S_i, chosen similar to those given in [4]. The shape function matrix is built from the shape functions in the following way $\mathbf{S} = [S_1\mathbf{I}\; S_2\mathbf{I}\; S_3\mathbf{I}\; S_4\mathbf{I}\; S_5\mathbf{I}\; S_6\mathbf{I}]$, in which $\mathbf{I}$ is the 2×2 identity matrix. The position vector $\mathbf{r}$ in (1) can be rewritten as

$$\mathbf{r} = \mathbf{r}_0 + \mathbf{u} = \mathbf{S}(\mathbf{q}_0 + \mathbf{q}). \tag{3}$$

3 Work of elastic forces and piezo-electric actuation

The goal of the present section is to formulate the virtual work for the presented element in "Geometric Description and Definition of the Finite Element", which not only shows axial, bending, shear and thickness deformation, but also includes piezo-electric components to realize the desired actuation.

The original equation for a shear deformable ANCF beam is based on the work of Omar and Shabana [8]. Herein, the energy of elastic forces follows from the continuum mechanics approach, using the relation between the nonlinear Green-Lagrange strain tensor and the second Piola-Kirchhoff stress tensor. In a recent study, [2], the continuum mechanics based approach is coupled with piezo-electric strains. The main focus of this paper is a formulation based on Reissner, Simo and Vu-Quoc, which represents a different derivation for the strain energy in terms of the axial strain Γ_1, the shear strain Γ_2 and the bending strain K. This formulation is coupled with piezo-electric components according to [13].

Reissner [9] stated the virtual work of internal forces as

$$\delta W_{int}^{Reissner} = \int_{-L/2}^{L/2} (M\delta K + N\delta\Gamma_1 + Q\delta\Gamma_2) d\xi. \tag{4}$$

The stress resultants, more precise the normal force N, the shear force Q and the bending moment M depend on a set of generalized strain measures, Γ_1, Γ_2 and K. In the present paper, a linear relation between stress resultants and generalized strains is used. The stress resultants and generalized strains are defined according to Simo and Vu-Quoc [12] and the different quantities are decoupled, setting

$$N = EA\Gamma_1, \qquad Q = k_s GA\Gamma_2, \qquad M = EIK. \tag{5}$$

For the above definitions, the axial stiffness EA, the shear stiffness GA with the shear correction factor $k_s = (10(1+\nu))/(12+11\nu)$ and the bending stiffness EI are used. Furthermore, E represents Young's modulus, A is the area of the cross section, G is the shear modulus and I is the second moment of area of the cross section. The quantity K denotes the change of angle of the cross section relatively to the reference length $K = (\mathbf{r}_{,\eta} \times \mathbf{r}'_{,\eta})/|\mathbf{r}_{,\eta}|^2$, while Γ_1 and Γ_2 are defined as $\Gamma_1 = \mathbf{t}_1^T\mathbf{r}' - 1$, $\Gamma_2 = \mathbf{t}_2^T\mathbf{r}'$, in which the vector $\mathbf{t}_2$ is introduced as $\mathbf{t}_2 = \frac{\mathbf{r}_{,\eta}}{|\mathbf{r}_{,\eta}|}$ and $\mathbf{t}_1$ is perpendicular to $\mathbf{t}_2$. The combination of (4) and (5) corresponds to the following weak form

$$\delta W_{int}^{Reissner} = \int_{-L/2}^{L/2} (EA\Gamma_1\delta\Gamma_1 + k_s GA\Gamma_2\delta\Gamma_2 + EIK\delta K)\, d\xi. \tag{6}$$

In order to include piezo-electric actuation in the ANCF beam element, the virtual work of internal forces has to be extended by the actuation of eigenstrains representing normal, shear and bending strain in the following way,

$$\delta W = \int_{-L/2}^{L/2} ((EA\Gamma_1 - N_a)\delta\Gamma_1 + (k_s GA\Gamma_2 - Q_a)\delta\Gamma_2 + (EIK - M_a)\delta K)\, d\xi, \tag{7}$$

in which N_a, M_a and Q_a are the actuating force, moment and transverse force, respectively. For the definition of these quantities, the axial eigenstrain representing the converse piezo-electric effect of thin piezo-electric layers according to [13] is used as $\varepsilon^0_{xx} = d_{31}E_z$, including the piezo-electric coefficient d_{31} and the electric field E_z in thickness direction of an actuating layer, all other eigenstrain components are zero. Hence, the term Q_a vanishes, see also [3]. Due to the anti-symmetric actuation of the piezo-electric patches, the term N_a is neglected for moderately large deformations. The actuating force and moment are defined as

$$N_a = \int_A E\varepsilon^0_{xx} dA \qquad \text{and} \qquad M_a = \int_A Ez\varepsilon^0_{xx} dA. \tag{8}$$

An additional term accounting for the strain energy due to thickness deformation has to be introduced, see [1]. The additional thickness strain energy $W^{thickness}$ can be defined, in case of a rectangular cross section and neglecting the influence of piezo-electric actuation, by

$$W^{thickness} = \frac{1}{2}\int_{-L/2}^{L/2} EAE^2_{\eta\eta} d\xi, \tag{9}$$

in which $E_{\eta\eta} = \frac{1}{2}(\mathbf{r}^T_{,\eta}\mathbf{r}_{,\eta} - 1)$ is the transverse component of the Green-Lagrange strain tensor. The quantity $E_{\eta\eta}$ is independent of the stress resultants Γ_1, Γ_2 and K defined above. Hence, the total virtual work is the sum of (7) and the variation of (9).

4 Numerical example: a highly flexible two-link robot

In order to validate the proposed ANCF element and to illustrate the performance, several static and dynamic examples taken from [1] have been investigated, see [5]. A convergence study showed the accuracy of the proposed element. The convergence rate of the nodal displacement follows $O(n^4)$ in the provided examples.

The quadratic ANCF actuating piezo-electric element has been implemented in the framework of the multibody and finite element research code HOTINT (see http://tmech.mechatronik.uni-linz.ac.at/staff/gerstmayr/hotint.html) in order to investigate the following numerical example. The two arms of a two-link robot are discretized with the help of the presented ANCF finite element in order to compute the tip displacements and analyze the following test setting.

A highly flexible two-link robot with tip masses is studied, as shown in Fig. 3. The two arms are modelled as beams with equal length $L_1 = L_2 = 0.5\,\text{m}$, cross

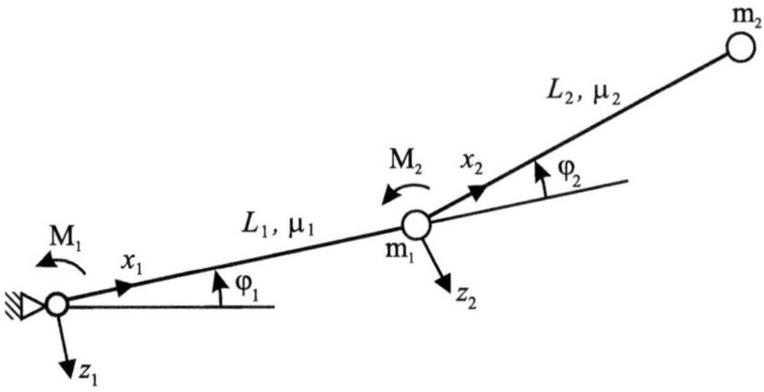

Fig. 3 Piezo-electrically actuated two-link robot

section of height $h = 0.01\,\text{m}$ and width $w = 0.02\,\text{m}$. The arms are made of a linear elastic material with Young's modulus $E = 7 \cdot 10^{10}\,\text{N/m}^2$, Poisson's ratio $\nu = 0.3$ and density $\rho = 2700\,\text{kg/m}^3$. The tip masses at the end of the arms are set to $m_1 = m_2 = 1\,\text{kg}$. Assuming that the motion takes place in a horizontal plane, gravity can be neglected. Additionally, the influence of the patches to the bending stiffness EI in (7) is neglected. On each arm three piezo patches are applied with length $l_p = 0.1\,\text{m}$ and with a length of the space between the patches of $l_s = 0.05\,\text{m}$. For a sketch of the realisation with piezo-electric patches see [13]. The shear correction factor in (7) is chosen as $k_s = 0.85$. The arms are driven by two motors providing the joint moments M_1 and M_2 in the joints. The joint angles φ_1 and φ_2 follow exactly a common trajectory for industrial robots, according to [13], which gives a smooth motion of the robot. The effects of the electrical drives on the gears are neglected. The mechanical system is assumed as idealized and damping-free. In the present paper, a shape control based on discrete patches is applied to the idealized two-link robot. The numerical example shows the performance of using only a discrete instead of a continuous distribution of piezo-electric actuation. In the following investigation, the motion with and without control by means of piezo-electric actuation is compared. The tip displacement in y-direction is taken as a measure for the comparison, see Fig. 4. The desired final vertical tip displacement is 0.5 m, however without control an oscillation of about 40 mm remains. Considering the open loop shape control in the system, the oscillations show an amplitude of approximately 0.7 mm.

5 Conclusion

In the present paper, a quadratic ANCF shear deformable beam element for describing a nonlinear large deformation beam with bending, axial and shear deformation property is presented. The deformation energy according to Simo and Vu-Quoc [12] is coupled with piezo-electric strains, and enhanced by an additional

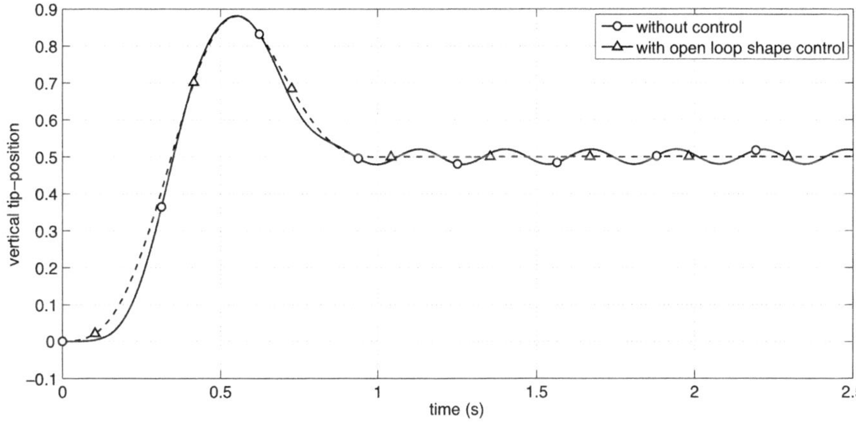

Fig. 4 Comparison of the vertical tip position of the robot without and with open loop control

term accounting for thickness deformation. A numerical model for an idealized flexible two-link robot is set up and combined with an open loop control algorithm, which allows the suppression of the vibrations of the robot arms.

Acknowledgements K. Nachbagauer acknowledges support from the Austrian Science Funds FWF via the project I337-N18, C. Zehetner and J. Gerstmayr from the K2-Comet Austrian Center of Competence in Mechatronics ACCM.

References

1. Gerstmayr, J., Matikainen, M., Mikkola, A.: A geometrically exact beam element based on the absolute nodal coordinate formulation. J. Multibody Syst. Dyn. **20**, 359–384 (2008)
2. Irschik, H., Gerstmayr, J.: A continuum mechanics approach for geometrically nonlinear shear-deformable and smart beams. In: Topping, B.H.V., Adam, J.M., Pallarés, F.J., Bru, R., Romero, M.L. (eds.) Proceedings of the Tenth International Conference on Engineering Computational Technology, p. 216. Civil-Comp Press, Stirlingshire, UK (2010)
3. Krommer, M., Irschik, H.: On the influence of the electric field on free transverse vibrations of smart beams. Smart Mater. Struct. **8**, 401–410 (1999)
4. Mikkola, A., Garcia-Vallejo, D., Escalona, J.L.: A new locking-free shear deformable finite element based on absolute nodal coordinates. Nonlinear Dyn. **50**, 249–264 (2007)
5. Nachbagauer, K., Gerstmayr, J., Pechstein, A., Irschik, H.: A linear and quadratic planar finite element based on the absolute nodal coordinate formulation. In: Proceedings of the First Joint International Conference on Multibody System Dynamics, Lappeenranta University of Technology, Finland (2010)
6. Nader, M.: Compensation of Vibrations in Smart Structures: Shape Control, Experimental Realization and Feedback Control, vol. 54. Reihe C, Naturwissenschaften und Technik, Trauner Verlag, Linz (2008)
7. Nader, M., Kaltenbacher, M., Krommer, M., von Garssen, H.-G., Lerch, R.: Active vibration control of a slender cantilever using distributed piezoelectric patches. In: Proceedings of the Thirteenth International Congress on Sound and Vibration Vienna (2006)

8. Omar, M., Shabana, A.: A two-dimensional shear deformable beam for large rotation and deformation problems. J. Sound Vibr. **243**(3), 565–576 (2001)
9. Reissner, E.: On one-dimensional finite-strain beam theory: the plane problem. J. Appl. Math. Phys. **23**, 795–804 (1972)
10. Shabana, A.: An absolute nodal coordinate formulation for the large rotation and deformation analysis of flexible bodies. Technical Report No MBS96-1-UIC, University of Illinois at Chicago, Chicago (1996)
11. Shabana, A.: Flexible multibody dynamics: review of past and recent developments. Multibody Syst. Dyn. **1**, 189–222 (1997)
12. Simo, J., Vu-Quoc, L.: On the dynamics of flexible beams under large overall motions – the plane case: Part I and II. J. Appl. Math. **53**, 849–863 (1986)
13. Zehetner, C., Gerstmayr, J.: Control of flexible vibrations in a two-link robot by piezoelectric actuation. In: Troch, I., Breitenecker, F. (eds.) Proceedings of the 6th Vienna Conference on Mathematical Modelling, 988–998 (2009)

Model Based Control of Heating Ventilating and Air-Conditioning Components

Jakob Rehrl and Martin Horn

Abstract Model based control strategies are not well established in the heating ventilating and air conditioning industry yet. However, these techniques offer considerable advantages over traditional approaches based on expert knowledge and therefore may overcome deficiencies of existing control concepts. This paper outlines a model based control policy for an industrial test plant. The method uses describing functions to avoid undesired limit cycles. The proposed technique proves to be a systematic tool to improve the performance of heating ventilating and air conditioning systems.

1 Introduction

The main objectives of heating ventilating and air conditioning (HVAC) systems are temperature and humidity control. In [4] the problem of limit cycles in temperature control systems is analyzed. For an air conditioning system of an office room, the describing function method is used to eliminate periodic oscillations of the room air temperature. Simulations reveal that the proposed approach is suitable for improving an existing PI-controller. In this article an industrial HVAC pilot plant is introduced in order to verify these simulation results in practice.

The paper is organized as follows: In "Test Plant" a mathematical description of the test plant and of its components is outlined. "Describing Function Method" is dedicated to the describing function method, "Conclusion" concludes the manuscript.

J. Rehrl (✉) · M. Horn
Institute for Smart System-Technologies, Control and Measurement Systems Group, Klagenfurt University, Austria
e-mail: jakob.rehrl@uni-klu.ac.at; martin.horn@uni-klu.ac.at

H. Irschik et al. (eds.), *Advanced Dynamics and Model-Based Control of Structures and Machines*, DOI 10.1007/978-3-7091-0797-3_21,

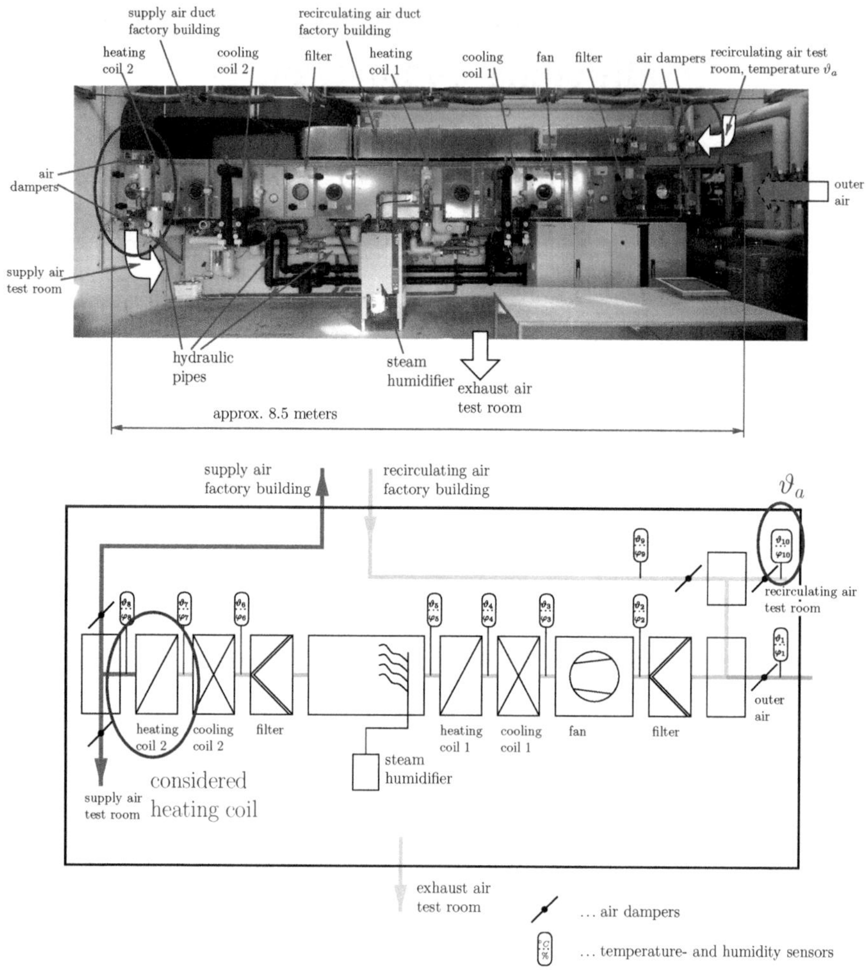

Fig. 1 Test plant

2 Test plant

In the following subsections, a functional description as well as a mathematical model of the pilot plant shown in Fig. 1 are given. The pilot plant was erected by the company Fischer&Co.[1] It serves for research purposes exclusively and offers the opportunities of

- Heating / cooling
- Humidifying / dehumidifying

air which can be transported to a factory building and/or to a test room.

[1]http://www.fischer-co.at.

2.1 Description of the test plant

A photo and a schematic representation of the test plant are shown in Fig. 1. Solely standard industrial HVAC components are integrated into the plant in order to provide realistic testing conditions. A CPU from the company Bernecker&Rainer[2] is used as a programmable control unit.

The major components are labeled in Fig. 1. The air passes the plant from right to the left. Via air dampers, the mixing ratio of outer air and air from the test room ($\approx 150\,\mathrm{m}^3$) or the factory building ($\approx 5000\,\mathrm{m}^3$) can be adjusted. A fan is installed to transport the air. The heating and cooling coils are installed to increase and to decrease the air temperature. The air humidity can be adjusted with the help of a steam humidifier. The air dampers on the left hand side of the test plant are used to select whether the test room or the factory building is supplied with conditioned air. In the present application, the room air temperature ϑ_a (see Fig. 1) is controlled using the heating coil (see "Describing Function Method").

In order to preserve transparency, only the utilized components, i.e. heating coil and room, will be explained in detail.

The used heating coil is a finned tube type heat exchanger. A fluid with a higher temperature than the air passing the heating coil increases the air temperature. With the help of a hydraulic circuit, the fluid temperature of the heating coil and consequently the heating power can be adjusted.

In the following experiments, the test room is supplied with conditioned air. The temperature ϑ_a of the air leaving the test room is the controlled variable.

2.2 Model of the test plant

In this section mathematical models of the HVAC components required for the application of the describing function method and for controller design are presented.

2.2.1 Heating coil

The mathematical model of the heating coil is obtained from mass and energy balances. The control input to the model is the water inlet temperature ${}^i\vartheta_w$. Further inputs which are assumed to be constant are the air inlet temperature ${}^i\vartheta_a$, the air mass flow $\dot{m}_a$ and the water mass flow $\dot{m}_w$. The model outputs are the outlet temperatures of air ${}^o\vartheta_a$ and water ${}^o\vartheta_w$. The tubes of the heating coil are divided into segments, see Fig. 2. In (1–5), temperatures are denoted by ϑ, massflows by $\dot{m}$. The subscripts w, a, i, o and t denote "water", "air", "inlet", "outlet" and "tube" respectively. For one segment, the following set of first order differential equations

[2]http://www.br-automation.com.

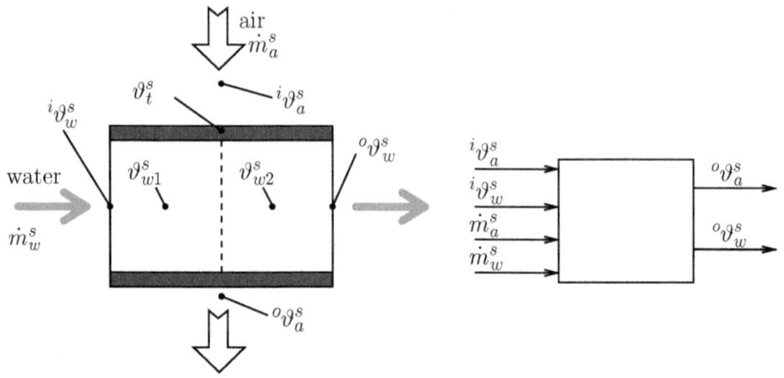

Fig. 2 One segment of the tubes of the heating coil

can be derived [4]:

$$\frac{d\vartheta_t^s}{dt} = k_1\left[0.5(\vartheta_{w2}^s + \vartheta_{w1}^s) - \vartheta_t^s\right] + k_2({}^i\vartheta_a^s - \vartheta_t^s) \tag{1}$$

$$\frac{d\vartheta_{w1}^s}{dt} = k_3\left[{}^i\vartheta_w^s - 0.5(\vartheta_{w1}^s + \vartheta_{w2}^s)\right] + k_4(\vartheta_t^s - \vartheta_{w1}^s) \tag{2}$$

$$\frac{d\vartheta_{w2}^s}{dt} = k_3(\vartheta_{w1}^s - \vartheta_{w2}^s) + k_4(\vartheta_t^s - \vartheta_{w2}^s) \tag{3}$$

The outputs can be computed via the algebraic relations:

$${}^o\vartheta_w^s = 1.5\,\vartheta_{w2}^s - 0.5\,\vartheta_{w1}^s \tag{4}$$

$${}^o\vartheta_a^s = e^{-\kappa}\,{}^i\vartheta_a^s + (1 - e^{-\kappa})\vartheta_t^s \tag{5}$$

Details concerning the parameters k_1 to k_4 and κ can be found in [5]. The model of the heating coil is found by interconnecting the segments. In the present case ten segments were used to model the heating coil.

The relation between mixing ratio of the fluids and the valve gear position was obtained via measurements.

2.2.2 Room

Due to a large number of influencing factors, rooms are extremely difficult to model. Therefore the room is represented by a first order element with dead time as suggested in [3]:

$$\frac{{}^o\vartheta_a^r(s)}{{}^i\vartheta_a^r(s)} \approx \left(K\frac{1}{1 + s\,T^r}\right)\cdot e^{-s\,T_d^r} \tag{6}$$

Note that ${}^o\vartheta_a^r$ and ${}^i\vartheta_a^r$ denote room air inlet and outlet temperature respectively. The gain K, the time constant T^r and the dead time T_d^r were identified from measured data.

3 Describing function method

The basic idea of the describing function method is simple, see e.g. [2]. Though it is a powerful technique to predict oscillations in nonlinear feedback loops. Figure 3 shows the structure of the nonlinear temperature control loop, where $R(j\omega)$ denotes the frequency response of the controller to be designed. $N(A,\omega)$ is the describing function which is composed of valve gear, valve, hydraulic circuit, heating coil and room. As usual, A and ω denote the amplitude and the frequency of a potential limit cycle

$$\Delta h \approx A \cdot \sin \omega t, \tag{7}$$

where Δh is the deviation of the valve gear position h from its (constant) operating point. As indicated in Fig. 3, it is the key issue to find intersection points of the describing surface and the controller frequency response. In case of at least one intersection point, limit cycles are likely to occur, otherwise no limit cycle is expected. Figure 4 shows the describing surface for the pilot plant operated in the vicinity of an operating point of $\vartheta_a = 17.5°\text{C}$ and a valve gear position h of 50%. The algorithm to compute $N(A,\omega)$ is outlined in [4].

3.1 *Controller design*

The starting point is a predefined standard PI-controller $R_1(s)$ with fixed controller parameters. The practical application of this controller in the pilot plant yields unacceptable behaviour. Instead of a desired steady state temperature of 17.5°C, the room air temperature ϑ_a shows an undesired oscillation with an amplitude of approximately 1.5°C, see Fig. 5. This poor performance is confirmed by the describing function method, i.e. the frequency response $R_1(j\omega)$ intersects the describing surface, see Fig. 4.

Based on the linear algebraic method [1] which is applied to a linearized plant model $P(s)$, an alternative controller $R_2(s)$ is computed. Specifications for the closed loop system are expressed in terms of maximum peak and rise time

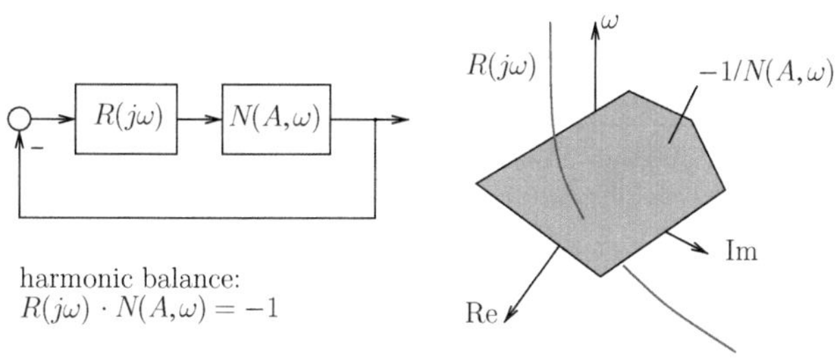

Fig. 3 Harmonic balance

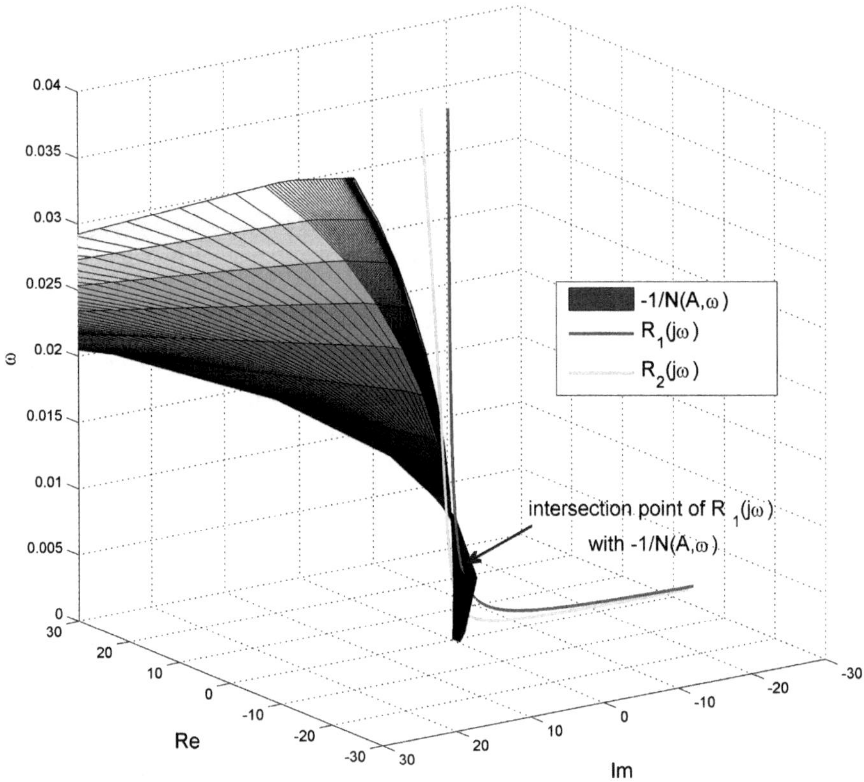

Fig. 4 Describing surface

$$M_p = 1.05 \qquad t_r = 150\,s \tag{8}$$

of the step response. This yields a desired closed loop transfer function

$$T(s) = \frac{0.00025}{s^2 + 0.02071\,s + 0.00025} \tag{9}$$

and finally the sought controller transfer function[3]

$$R_2(s) = \frac{1.3881(s^2 + 0.01559s + 6.519 \cdot 10^{-5})(s^2 + 0.1466s + 0.008878)}{s(s + 0.1047)(s^2 + 0.03189s + 0.002053)}. \tag{10}$$

[3]In order to achieve the desired tracking performance, a prefilter $V(s)$ is also included. However, it has no impact on the limit cycle analysis and is therefore omitted.

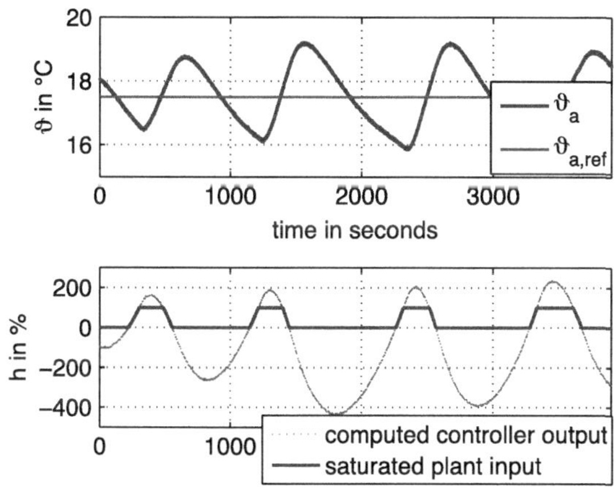

Fig. 5 Experimental result using controller $R_1(s)$

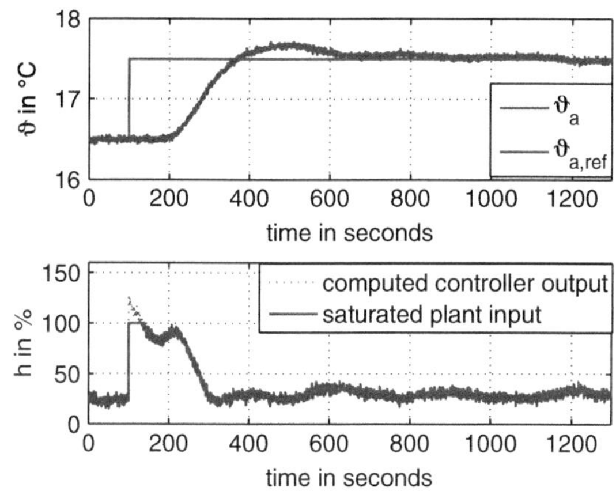

Fig. 6 Experimental result using controller $R_2(s)$

Note that the design procedure is an iterative process. If the frequency response $R_2(j\omega)$ intersects $-1/N(A,\omega)$, the design has to be repeated with more conservative specifications. The frequency response of the final controller $R_2(s)$ has no intersection point with the describing surface (see Fig. 4), consequently it is unlikely that limit cycles will occur. Experimental results shown in Fig. 6 confirm the absence of limit cycles and reveal very satisfactory overall performance of the feedback loop.

4 Conclusion

In this article, the implementation of a previously published design procedure on a real world HVAC system is outlined. The approach which is based on the describing function method and the linear algebraic method proves to be well suited for HVAC applications facing the problem of limit cycles.

Acknowledgements This project is partially funded by the Austrian Research Promotion Agency (FFG), project number 820141, project title "Hochgenaue Regelung klimatechnischer Anlagen".

References

1. Chen, C.T.: Analog and Digital Control System Design: Transfer-Function, State-Space, and Algebraic Methods. Saunders College Publishing, Orlando, FL (1993)
2. Khalil, H.K.: Nonlinear Systems, 3rd edn. Prentice Hall, Englewood Cliffs, NJ, USA (2001)
3. Rehrl, J., Horn, M.: Analysis and synthesis of a temperature control system using describing functions. at-Automatisierungstechnik **56**(6), 296–306 (2008), DOI 10.1524/auto.2008.0709
4. Rehrl, J., Horn, M., Reichhartinger, M.: Elimination of limit cycles in hvac systems using the describing function method. In: Decision and Control, 2009 held jointly with the 2009 28th Chinese Control Conference. CDC/CCC 2009. Proceedings of the 48th IEEE Conference on Decision and Control and 28th Chinese Control Conference, pp. 133–139 (2009), DOI 10.1109/CDC.2009.5400857
5. Wiening, W.: Zur Modellbildung, Regelung und Steuerung von Wärmeübertragern zum Heizen und Kühlen von Luft. Fortschritt-Berichte VDI Reihe 8 Nr. 128, VDI-Verlag, Düsseldorf (1987)

Microsensors Based on Mechanically Vibrating Structures

Erwin K. Reichel and Bernhard Jakoby

Abstract The frequency response of electromechanical resonators is sensitive to the material properties of a surrounding medium and therefore can be used as sensors for viscoelastic properties and density. A good theoretical understanding of the fluid-structure interaction is necessary for a proper interpretation of the measurements and to guide sensor design and optimization. We present the sensor principle and focus on resonators using electrodynamic transduction and can be fabricated by technologies that allow mass production. Exemplarily, two sensor geometries are discussed. In one case, the top and bottom of the sample volume are actuated. In the other case, a metallic resonator element is immersed in the sample liquid, realizing a dominant shear-wave interaction.

1 Introduction

Sensors based on mechanical resonators are spreading into more and more fields of applications and hence increase in diversity. It is not easy to date back the first of its kind but one of first widespread application in high-tech industry was the quartz crystal microbalance (QCM) used to weigh thin films with unprecedented accuracy [1]. Later it was found that their (the thickness shear mode – TSM – type, to be more specific) operation in liquid media allows for an elegant way to measure the viscosity of Newtonian liquids [2], and even the high-frequency behavior of viscoelastic layers [3–5]. Functionalized surfaces have led to a variety of sensors

E.K. Reichel (✉)
Centre for Surface Chemistry and Catalysis (COK), KU Leuven, 3001 Leuven, Belgium
e-mail: er@ieee.org

B. Jakoby
Institute for Microelectronics and Microsensors, JKU Linz, Austria
e-mail: bernhard.jakoby@jku.at

H. Irschik et al. (eds.), *Advanced Dynamics and Model-Based Control of Structures and Machines*, DOI 10.1007/978-3-7091-0797-3_22,

for chemical sorption [6], cell growth [7], and other biochemical applications. The detailed analysis of operation modes and interaction mechanisms is still an active field of research [8, 9]. Another principle based on the piezoelectric effect are surface–guided shear wave transducers operating at very high frequencies (above 100 MHz) [10].

Whereas density is a static property depending on the mass and number of particles per volume, viscosity depends on intermolecular processes at various time–scales and intermolecular forces and structure at wide-ranged length scales. Not surprisingly, this fact and emergence of new materials, e.g. polymeric and micellular solutions, manifold dispersions, has led to the establishment of rheology as a distinct subject of science.

With resonator methods, the medium is probed for its linear viscoelastic response in a limited frequency range. The deformation is wave-like, so the density contributes to the sensor response. An ideal viscosity sensor mimics the operation of a laboratory rheometer while retaining its online capabilities. Besides TSM quartz and surface-guided shear waves, vibrating wires, torsional resonators, or tuning forks are employed to derive viscosity from the resonant frequency response – usually the damping or the resonance frequency shift compared to the unloaded case or both are evaluated.

A device that becomes very popular for microsensors, but originated from atomic force microscopy (AFM) is the micromachined cantilever. Regularly, special tips or functional coatings provide the mechanism for sensing in gaseous or liquid media. The latter requires a model to incorporate the fluid-structure interaction and concise signal processing to obtain reliable sensor readings [11]. More sophisticated micromachined resonators are employed in MEMS gyroscopes, integrated frequency standards, filtering, and sensors for density [12] and viscosity [13].

Here we focus on miniaturized systems with online capability for measuring rheological properties (viscosity at elevated frequencies) and density of liquids. Electrodynamic actuation (Lorentz forces) is used to achieve significant vibration amplitudes while refraining from delicate (piezoelectric) materials. The devices are designed to have integrated readout mechanisms such as inductive or piezoresistive vibration measurement.

2 Viscoelastic material behavior

Rheological constitutive models of materials are described by the deformation response caused by applied forces or the reciprocal. Transitions in the microstructure regularly lead to behavior which is neither captured by the linear elastic (Hookean) body nor the simple Newtonian liquid. Even though the topic is widespread and a large variety of methods to measure specific properties have been developed, we will concentrate on a domain known as linear viscoelasticity. It can be viewed as the combination of the two basic material models mentioned before. Studying the viscoelastic response can reveal a surprisingly large amount of information about constitutive processes at the microscale [14], and consequently is an established method in polymer science and processing, as well as the characterization of suspensions and emulsions.

To outline the underlying basic models, the idealized shear experiment is considered, where two plates are slid over another with the material to be studied in between. The process is assumed to be stationary, so that a linear deformation profile is established. Additionally, inertial and boundary effects are neglected. In the case that the material between the plates is an elastic solid, the force is proportional to the displacement of the upper plate, hence the constitutive equation is Hooke's law,

$$\tau = G'\gamma, \tag{1}$$

where τ is the shear stress, G' the elastic modulus, and γ the shear deformation. As soon as the force vanishes, the plate slides back to its initial position, therefore G' is also called storage modulus. Now consider that the gap is filled with a liquid. Newton proposed that the force is now proportional to the displacement speed, or the stress is proportional to the shear deformation rate,

$$\tau = \eta\dot{\gamma}, \tag{2}$$

where η is the viscosity, and $\dot{\gamma}$ is the time derivative of the deformation. When the force is switched off, the plate will stop moving and remain in its position. The power required to maintain the movement has been dissipated.

As we consider linear phenomena only, the introduction of complex notation to describe harmonic oscillation is near at hand. Accordingly, Eqn. 2 is written as

$$\tau = \mathrm{j}\omega\eta\gamma, \tag{3}$$

where ω is the angular frequency of oscillation (omitted is the common factor of $\mathrm{e}^{\mathrm{j}\omega t}$ and taking the real part).

Now there are two ways of combining these basic constitutive models, in analogy to serial and parallel connection of electrical components. One can add the force models to obtain what is known as the Kelvin-Voigt body,

$$\tau = (G + \mathrm{j}\omega\eta)\gamma, \tag{4}$$

or add the deformation,

$$\gamma = \left(\frac{1}{G} + \frac{1}{\mathrm{j}\omega\eta}\right)\tau \quad \text{or} \quad \tau = \frac{1}{1/G + 1/(\mathrm{j}\omega\eta)}\gamma \tag{5}$$

to obtain Maxwell's fluid model. The distinction between solid and liquid can be made in the terminal region, where ω tends to zero. The Maxwell fluid reduces to the Newtonian behavior, whereas the Kelvin-Voigt body reduces to the Hookean model. Both of the introduced viscoelastic models are seldom encountered for real materials, but multiple elements are used to describe the response in a wide frequency range [15].

In a frequency sweep oscillatory measurement, the quantity derived from the measurement of force amplitude and phase (with respect to the deformation) is the complex modulus $G^* = G' + \mathrm{j}G''$, where G' is called storage modulus, and G'' loss modulus. Another way is to quantify the complex viscosity $\eta^* = \eta' + \mathrm{j}\eta''$.

There exists a certain amount of freedom in the model parameters, so that with $T_\mathrm{r} = \eta/G$, the complex viscosity of a Maxwell fluid is sometimes written as

$$\eta^* = \frac{\eta_0}{1 + \mathrm{j}\omega T_\mathrm{r}}, \tag{6}$$

with zero-shear viscosity η_0 and relaxation time constant T_r. This description accounts for the phenomenon of stress relaxation by a diffusive process. If the deformation process is faster than the relaxation process, the material behaves more like a solid. In the case of slow deformation, the medium is liquid-like. The material appears to have a kind of memory with a certain decay time.

In the ideal case, the viscoelastic response over several orders of magnitude in frequency is measured. Conventional rheometers span a range from about 0.01–100 Hz, limited by the eigenfrequency of the measurement system and wave phenomena in the gap.

Effects that are not captured by the linear viscoelastic model are frequently encountered. Dependence of the material parameters on the amplitude of deformation or deformation rate would cause higher harmonics in the measured response. What can be measured with modern rheometers are stresses perpendicular to the direction of shear. These normal stresses are second order effects, so they are observed at the double frequency. Time dependence is regularly observed when the material undergoes permanent structural changes.

3 Sensors

State-of-the-art sensors for rheological properties rely on the resonance principle. A mechanical structure with at least one defined resonance mode is brought in contact with the medium under test. The reaction forces cause changes in the mechanical frequency response, which is transformed into an electrical signal. In most cases, a good signal-to-noise ratio (SNR) can be achieved only in the vicinity of the resonance frequency of the employed eigenmode. This is the major difference to that used in conventional, force-transduction methods, as utilized in rotational or oscillatory rheometers.

Compared to the unloaded case, the resonance bandwidth is increased and the peak is shifted to lower frequencies by an amount depending on the density and viscoelastic properties at the respective frequency. The accurate modeling of the fluid-structure interaction is a rather complex task because the deformation field surrounding the resonator element is wave-like, and hence dependent on the properties to be measured. Therefore, the relation between measured parameters and fluid properties is merely empirical and often limited to the measurement of Newtonian liquids.

In the following, a few selected sensor principles investigated by the authors and coworkers are presented. The aim of the ongoing research is to achieve better theoretical understanding, establish novel sensor principles, and use state-of-the-art fabrication techniques to increase the range of application for online measurement of rheological properties. A focus is set to technologies that allow mass production, e.g. lithography, polymer processing techniques or silicon micromachining at wafer scale.

In Fig. 1 an overview of the investigated concepts is shown. Concerning the fluid-structure interaction, the important design parameters are the operating (resonance) frequency, the ratio of in-plane to out-of-plane moving surfaces in contact with the liquid, and the vibration amplitude. The penetration depth of the velocity field, the achievable quality factors and SNR, and the applicability to complex structured liquids follow from these design parameters.

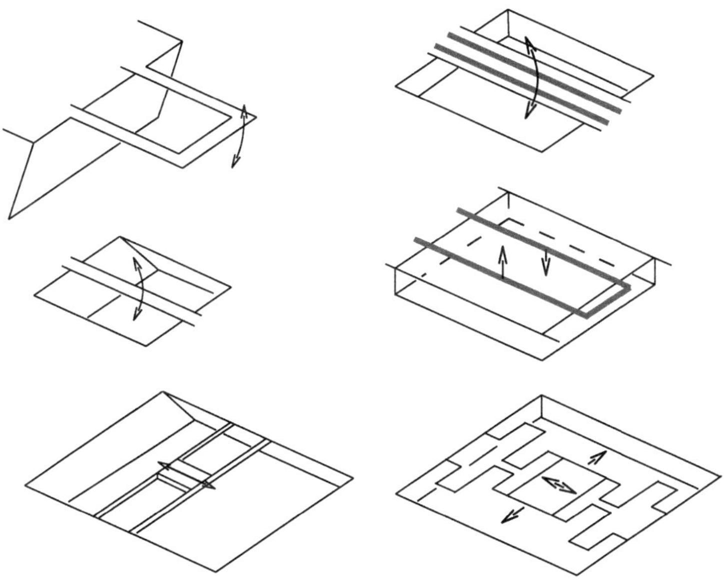

(**a**) micromachined (**b**) millimeter-sized

Fig. 1 Overview of the investigated resonators for viscosity and density analysis

3.1 Resonator modeling

The sensors considered here exhibit small vibration amplitudes, so that the constitutive (material) relation is assumed to be linear. This also implies that convective motion is negligible [16]. The resonator has a mechanism to generate a force from an electrical signal and some kind of transduction mechanism to generate an electrical signal proportional to the deformation. While piezoelectric resonators generate stress from electrical voltage and charge from deformation, the Lorentz-force resonators presented here generate stress from electrical currents in a constant magnetic field and an induction voltage proportional to the displacement velocity. In either case, the resonance frequency and damping is dominantly defined by the mechanical properties of the transducer and the surrounding medium. Spurious electrical coupling often affects the actually measured signal and needs to be separated properly.

The aim of the resonator modeling is to derive a description of the coupled system in the form of the frequency response of the input/output relation as a function of the known sensor characteristics and the fluid parameters to be measured. This relation then is inverted for the interpretation of the sensor readout.

The stress tensor $\boldsymbol{T}$ is written as a function of the strain tensor $\boldsymbol{S}$,

$$\boldsymbol{T} = \underline{\underline{c}} : \boldsymbol{S} + \underline{\underline{\eta}} : \frac{\partial}{\partial t}\boldsymbol{S} = \left(\underline{\underline{c}} + \mathrm{j}\omega\underline{\underline{\eta}}\right) : \boldsymbol{S}, \tag{7}$$

where time-harmonic complex notation has been used in the latter form, and $\underline{\underline{c}}$ is the 6×6 stiffness matrix and $\underline{\underline{\eta}}$ the viscosity matrix. In the general, viscoelastic case, both $\underline{\underline{c}}$ and $\underline{\underline{\eta}}$ are frequency dependent and related to each other by Kramer–Kronig properties [15]. In liquids and soft solids, incompressibility is a valid assumption in the moderate frequency range considered here [17]. Only isotropic materials are considered here. In this case only the components which relate the shear deformation to the shear stress denoted as $c_{44} = c_{55} = c_{66}$ and $\eta_{44} = \eta_{55} = \eta_{66}$ enter the governing equations.

Although the resonator is a distributed system, the sensor response is described as a function of a few parameters only, which requires a reduction in the modeling order. Resonators are commonly modeled by a decomposition in eigenmodes. The displacement field, i.e. the solution of the equation of motion, is written as a series,

$$\mathbf{u}(\boldsymbol{r}, t) \approx \sum_{i=1}^{N} q_i(t)\boldsymbol{U}_i(\boldsymbol{r}), \tag{8}$$

where $q_i(t)$ are the reduced or generalized coordinates. $\boldsymbol{U}_i(\boldsymbol{r})$ are the non-trivial solutions (free modes of oscillation at angular frequency ω) of the mechanical system, inserting the time-harmonic steady state solution $\mathbf{u}(\boldsymbol{r}, t) = \boldsymbol{U}(\boldsymbol{r})\mathrm{e}^{\mathrm{j}\omega t}$,

$$\nabla \cdot \left[\underline{\underline{c}} : \nabla_{\mathrm{s}}\boldsymbol{U}_i(\boldsymbol{r})\right] + \rho\omega_i^2\boldsymbol{U}_i(\boldsymbol{r}) = \boldsymbol{0}. \tag{9}$$

The calculation domain is restricted by appropriate boundary conditions. Each of these eigenmodes corresponds to an eigenfrequency ω_i (also referred to as the undamped resonance frequency as we did not consider $\underline{\underline{\eta}}$ in the evaluation). The advantage of this approach is that only the eigenmodes with significant contribution to the observed frequency range, often only one up to a few N, have to be considered to obtain a reasonable approximation.

Except for basic cases it is not possible to find an analytic expression for the eigenmodes, but approximation techniques are available, e.g. Ritz ansatz or finite element analysis (FEA) [18]. The latter is very effective even for complex geometries, but numerically expensive in liquid or viscoelastic materials at elevated frequencies. The reason is that rapidly decaying shear waves demand a very fine resolution of discretization near the surface. A much more efficient way for the modeling of the fluid-interaction is to solve for the eigenmodes in the unloaded case and describe the resonator loading in terms of the resulting surface velocity. This concept comes close to boundary elements methods [17].

As long as the wavelength of compressional (acoustic) waves in the medium is sufficiently large compared to the structure itself, the fluid can be assumed to be incompressible [17]. In that case, the displacment fluid velocity field is solved by a Laplace equation (potential flow in an ideal fluid), where the boundary conditions match the normal velocity component. For a convex shape of the sensor element, the oscillation will excite decaying shear waves at the boundary. Using the plane wave solution of decaying waves in viscoelastic liquids as an approximation for the displacement fields in the vicinity of the surface, the power dissipated in the liquid can be obtained by integrating the real part of the complex acoustic Poynting vector [20] over the surface of the vibrating body [16].

Alternatively, the additionally stored (kinetic or elastic) energy as well as the dissipated power is derived from the velocity field in the liquid [19]. Knowing these energy terms, the quality factor (Q-factor) is obtained by inserting into the definition [21]

$$Q = \frac{\omega_0 \times \text{stored energy}}{\text{average power loss}}. \tag{10}$$

The challenge of the sensor design is to impose a deformation field for which the inversion of this dependence can be carried out efficiently and with the required accuracy. For quartz shear mode resonators, the approximation of a plane shear wave has been applied regularly, and the error made by this simplification can be estimated [8]. For resonators at sufficiently low oscillation frequency, most fluids behave incompressible. In that case, analytical expressions for cantilever-like structures can be found [22].

In the following, two sensor principles are presented exemplarily where the modeling of the sensor response was carried out. A fully numerical modeling, e.g., by using FEA is often not viable because the wave-like deformation profile in the liquid would require a discretization which is beyond the ability of standard simulation tools.

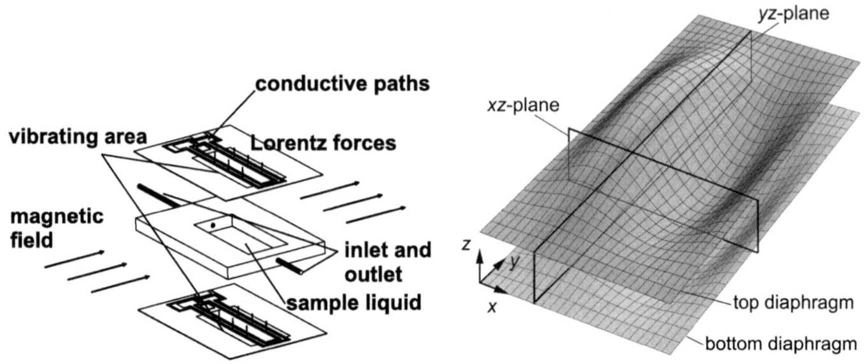

(**a**) schematic of the sensor cell (**b**) eigenmode of the membranes

Fig. 2 Double membrane cell

3.2 Examples of integrated sensors

A sensor fabricated by low-cost polymer technology is shown in Fig. 2 and described in detail in [19]. The liquid is injected with needles in the sample volume between the two actuated membranes. Conductive copper paths are used for excitation of the anti-symmetric membrane mode shown in Fig. 2b, and reciprocally to measure the vibration amplitude and phase by the motion-induced voltage. The frequency response in the vicinity of the resonance frequency corresponding to the respective eigenmode is measured by a frequency sweep of a sinusoidal signal. The induced voltage is in the range of microvolts, but can be measured with adequate SNR using a lock-in amplifier setup.

The deformation field between the two membranes depends on the velocity at the boundaries and the material properties, i.e., density and viscoelastic properties. It is assumed that the membrane mode shape is not altered by the fluid-structure interaction, so the energy dissipation and additional kinetic energy in the fluid can be calculated. Compressibility at the investigated frequency can be neglected. The solution of the velocity field can be split up in a potential function (Laplace equation) with the normal velocity as boundary condition, and decaying shear waves representing the rotational velocity component [19].

The second sensor element which is considered as an example consists of a 0.1 mm thick gold-coated Nickel-brass plate suspended on four spring structures, structured using lithography and wet-etching as shown in Fig. 3a, see [23]. A sinusoidal current is applied to two of the electrodes to excite the vibration by Lorentz-forces in combination with the perpendicularly oriented magnetic field. Reciprocally, an electric field is generated in the moving conductor. This motion-induced voltage is read out in a second electric loop via the remaining two electrodes.

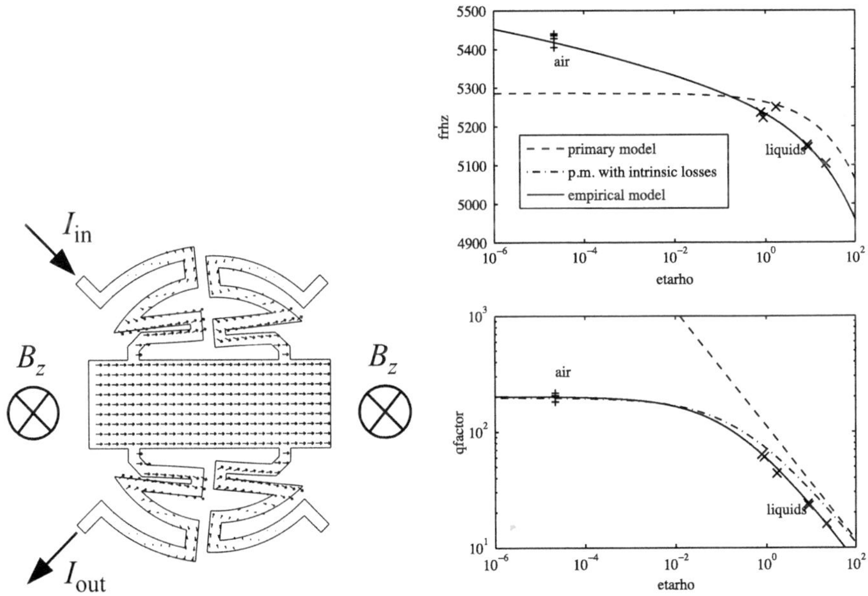

(**a**) metallic resonator element (**b**) modeling and measurements

Fig. 3 Metallic resonator sensor

The mechanical and electrical behavior of the device are modeled with FEA using COMSOL multiphysics, including an approximation of the fluid-structure interaction. The latter assumes that the in-plane oscillation of the sensor surface in contact with the liquid moving generates a plane shear wave penetrating into the liquid. We assume that the solution valid for an infinitely extended plane can be used to calculate the resistance force per area. This assumption is justified when the penetration depth of the shear wave (below 32 μm for the fluids used here) is small compared to the plate dimensions, so that boundary effects can be neglected. The actual sensor response shown in Fig. 3b shows a slightly deviating behavior. This result together with more detailed numerical analysis leads to the conclusion that boundary effects still are significant. To obtain a sensor calibration function, the relation between the measured resonance frequency and quality factor to viscosity and density of the test liquids was modified empirically [23].

4 Conclusions

Electromechanical resonators are used for sensing of viscoelastic material properties. The resonance frequency and Q-factor are sensitive to the viscoelastic shear modulus and the mass density of the surrounding fluid. The amount of dissipated energy is dependent on the penetration depth of the acoustic interaction, which is determined by the density as well as the elastic behavior of the medium. The modeling of the sensor response is a relatively complex task because different physical domains are involved and finite elements quickly become numerically expensive. FEA is used to calculate mode shape function for the corresponding resonance frequencies. Approximate methods are used to calculate the fluid structure interaction in order to derive a quantitative description of the sensor response. Exemplarily, two sensor geometries were discussed. In one case, the top and bottom of the sample volume are actuated. In the other case, a metallic resonator element is immersed in the sample liquid.

Acknowledgements This research represents an ongoing cooperation between the JKU Linz and KU Leuven. C.E.A. Kirschhock (COK, KU Leuven) and J. Vermant (CIT, KU Leuven) are acknowledged for support and discussions, Austrian Center for Competence in Mechatronics (ACCM), ESA and the Belgian PRODEX program for funding.

References

1. Sauerbrey, G.: Verwendung von Schwingquarzen zur Wägung dünner Schichten und zur Mikrowägung, Zeitschrift fr Physik A Hadrons and Nuclei 155/2, pp. 206–222 (1987)
2. Martin, S.J., Frye, G.C., Wessendorf, K.O.: Sensing liquid properties with thickness-shear mode resonators. Sens. Act. A: Phys. **44**, 209–219 (1994)
3. Reed, C., Kanazawa, K., Kaufman, J.: Physical description of a viscoelastically loaded at-cut quartz resonator. J. Appl. Phys. **68**, 1993–2001 (1990)
4. Lucklum, R., Behling, C., Cernosek, R.W., Martin, S.J.: Determination of complex shear modulus with thickness shear mode resonators. J. Phys. D: Appl. Phys. **30** (1997)
5. Johannsmann, D., Mathauer, K., Wegner, G., Knoll, W.: Viscoelastic properties of thin films probed with a quartz-crystal resonator. Phys. Rev. B **46**, 7808–7815 (1992)
6. Biemmi, E., Bein, T.: Assembly of nanozeolite monolayers on the gold substrates of piezoelectric sensors. Langmuir **24**, 11196–11202 (2008), doi:10.1021/la8009892
7. Heitmann, V., Rei, B., Wegener, J.: The quartz crystal microbalance in cell biology: basics and applications. Springer Series on Chemical Sensors and Biosensors, vol. 5, Part B, pp. 303–338. Springer-Verlag, Berlin, Heidelberg, New York (2007), DOI: 10.1007/5346_031
8. Beigelbeck, R., Jakoby, B.: A two-dimensional analysis of spurious compressional wave excitation by thickness-shear-mode resonators. J. Appl. Phys. **95**, 4989–4995 (2004), doi:10.1063/1.1697637
9. Ballato, A.: MEMS fluid viscosity sensor. IEEE Trans. UFFC **57**(3), 669–676 (2010)
10. Jakoby, B., Vellekoop, M.J.: Viscosity sensing using a love-wave device. Sensors Actuators A: Phys. **68**, 275–281 (1998)
11. Fadel, L., Lochon, F., Dufour, I., Francais, O.: Chemical sensing: millimeter size resonant microcantilever performance. J. Micromech. Microeng. **14**, S23 (2004)
12. Enoksson, P., Stemme, G., Stemme, E.: Fluid density sensor based on resonance vibration. Sensors and Actuators A: Physical. Elsevier, Amsterdam (1995)
13. Riesch, C., Reichel, E.K., Jachimowicz, A., Schalko, J., Hudek, P., Jakoby, B., Keplinger, F.: A suspended plate viscosity sensor featuring in-plane vibration and piezoresistive readout. J. Micromech. Microeng. **19**(7), 075010 (2009)
14. Larson, R.G.: The Structure and Rheology of Complex Fluids. Oxford University Press, Oxford (1999)
15. Macosko, C.W.: Rheology: Principles, Measurements, and Applications. Wiley, NY, USA (1994)

16. Landau, L.D., Lifshitz, E.M.: FluidMechanics. Pergamon Press, London (1959)
17. Wei, B., Reichel, E.K., Jakoby, B.: Modeling of a clamped–clamped beam vibrating in a fluid for viscosity and density sensing regarding compressibility. Sensors Actuators A: Phys. **143**, 293–301 (2008)
18. Weaver W., Jr, Timoshenko, S.P., Young, D.H.: Vibration Problems in Engineering, 5th edn. Wiley, NY, USA (1990)
19. Reichel, E.K., Riesch, C., Keplinger, F., Jakoby, B.: Modeling of the fluid-structure interaction in a fluidic sensor cell. Sens. Act. A: Phys. **156**(1), 222–228 (2009)
20. Auld, B.A.: Acoustic Fields and Waves in Solids, vol. 1. Wiley, NY, USA (1972)
21. Auld, B.A.: Acoustic Fields and Waves in Solids, vol. 2. Wiley, NY, USA (1972)
22. Eysden, C.A.V., Sader, J.E.: Small amplitude oscillations of a flexible thin blade in a viscous fluid: Exact analytical solution. Phys. Fluids **18**, 123102 (2006)
23. Reichel, E.K., Riesch, C., Keplinger, F., Jakoby, B.: Analysis and experimental verification of a metallic suspended plate resonator for viscosity sensing. Sensors Actuators A: Phys. **162**(2), 418–424 (2010)

Flatness Based Control of Linear and Nonlinear Systems

Kurt Schlacher, Markus Schöberl, and Martin Staudecker

Abstract The property of flatness has become quite popular especially for the class of lumped parameter systems since its introduction about 20 years ago. Challenging design problems like planning of optimal trajectories, stabilization of equilibria or trajectories by state feedback are significantly easier to solve for flat systems than for non flat ones. E.g. two point boundary problems, which appear in trajectory planning or in optimal system design, simplify, and the stabilization problem, formulated in flat coordinates, can be solved by linear methods. This contribution presents a short introduction into flatness and related control loop design problems for the class of nonlinear lumped parameter systems and applies the presented theory to the design of time optimal trajectories for a single mast stacker crane.

1 Introduction

Since the introduction of flatness for lumped parameter systems into control about 20 years ago, this approach has become quite popular especially for the class of lumped parameter systems. The responsible fact is that certain in general challenging design problems like planning of optimal trajectories, stabilization of equilibria or trajectories by state feedback are significantly simpler for flat systems than for non flat ones. Therefore, it is the aim of this contribution to present a short introduction into these control loop design problems, where we touch algebraic and differential geometric methods. But, we will not treat the problem to test a non

K. Schlacher (✉) · M. Schöberl
Institute of Automatic Control and Control Systems Technology, Johannes Kepler University, Linz, Austria,
e-mail: kurt.schlacher@jku.at; markus.schoeberl@jku.at

M. Staudecker
Bernecker + Rainer Industrie-Elektronik Ges.m.b.H., Linz, Austria
e-mail: martin.staudecker@br-automation.com

H. Irschik et al. (eds.), *Advanced Dynamics and Model-Based Control of Structures and Machines*, DOI 10.1007/978-3-7091-0797-3_23,

linear system with respect to flatness. The interested reader finds a comprehensive exposure of the theory for lumped parameter systems e.g. in [1], please see also the citations therein, the book [3] presents necessary and sufficient conditions for non linear lumped parameter systems to be flat, but these conditions are not applicable in a straightforward manner, a constructive approach for affine derivative systems based on Pfaffian systems can be found in [5]. It is also worth mentioning that flatness can be extended to distributed parameter systems, see e.g. [4].

According to the remarks from above this contribution is organized as follows. In "Motivation and some examples" we give a motivation for the introduction of flatness, where we start with potentials, well known from classical field theories. How the controller design problem can be significantly simplified for flat systems, is shown in "Control loop design". The design of time optimal trajectories for a single mast stacker crane is presented in "Time optimal trajectories for a single mast stacker crane".

2 Motivation and some examples

Flatness has some similarities with an old idea in physics, the theory of potentials. Let us consider the problem to characterize all force fields with the property

$$\int_\gamma \sum_{i=1}^n F_i \mathrm{d}x^i = 0,$$

where γ denotes a arbitrary closed path in $\mathbb{R}^n$. Let us assume the condition

$$\mathrm{d}P = \sum_{i=1}^n \partial_i P \mathrm{d}x^i = \sum_{i=1}^n F_i \mathrm{d}x^i, \tag{1}$$

or equivalently

$$\int_\gamma \mathrm{d}P = 0,$$

is met, then we have to check first, whether a solution of (1) exists. A necessary condition follows obviously from

$$\mathrm{d}\sum_{i=1}^n F_i \mathrm{d}x^i = \mathrm{d}\mathrm{d}P = 0, \tag{2}$$

and the Lemma of Poincaré tells us, when it is also sufficient, e.g. see [2]. If the problem is solvable, then a solution follows easily from

$$P(\gamma_1) - P(\gamma_0) = \int_\gamma \sum_{i=1}^n F_i \mathrm{d}x^i, \tag{3}$$

where γ is any path that joins the points $\gamma_1, \gamma_0 \in \mathbb{R}^n$ and γ_0 is a chosen point of reference. Let us summarize our observations, we have a problem like (1), a test for its solvability (2) and a formula (3) to derive a solution.

Differential geometry has been used to solve the problem to construct a potential for a force field. But there exists also an algebraic approach. Let the variable s_i be a shortcut for ∂_i and s for $s_1, \ldots, s_n$, then

$$F = M(s)\,P, \quad M^T = \left[s_1, \ldots, s_n \right], \quad F^T = \left[F_1, \ldots, F_n \right],$$

with $M(s) \in \mathbb{R}^{n\times 1}[s]$ is the counterpart to (1). The analogous problem to (2) is to construct a matrix $N(s) \in \mathbb{R}^{1\times n}[s]$, such that $NM = 0$ is met. This problem can be solved easily, but note, that it must be solved in the ring $\mathbb{R}[s]$ of polynomials with n unknowns s_i and real coefficients. It is straightforward to see that one finds $n(n-1)/2$ independent non trivial solutions.

Now let us turn to control problems and consider the simple system

$$\dot{x} = ax + bu\,. \tag{4}$$

Let y, y_1 denote a variable together with its first time derivative, or equivalently the assignment $y = f(t)$ implies $y_1 = \partial_t f(t)$. The system (4) is called flat, if we are able to construct a map, for linear and time invariant systems the choice

$$\begin{bmatrix} x \\ u \end{bmatrix} = T \begin{bmatrix} y \\ y_1 \end{bmatrix}, \quad T = \begin{bmatrix} t_{11} & t_{12} \\ t_{21} & t_{22} \end{bmatrix} \in \mathbb{R}^{2\times 2} \tag{5}$$

is sufficient, such that (4) is identically fulfilled. From

$$t_{11}y_1 + t_{12}y_2 = a\left(t_{11}y + t_{12}y_1\right) + b\left(t_{21}y + t_{22}y_1\right)$$

one derives easily the solution

$$T = \begin{bmatrix} b & 0 \\ -a & 1 \end{bmatrix}$$

provided $b \neq 0$ is met. In the case $b = 0$ the system is not flat. Also this problem can be solved by algebraic methods. To demonstrate this we rewrite (4) as

$$\left[(s-a) \; -b \right] \begin{bmatrix} x \\ u \end{bmatrix} = 0,$$

where s denotes the time derivation. Now, we look for an explicit representation for the solution of this equation, or equivalently, we try to solve the equation

$$\underbrace{\begin{bmatrix} (s-a) & -b \\ a_{21} & a_{22} \end{bmatrix}}_{M} \begin{bmatrix} x \\ u \end{bmatrix} = \begin{bmatrix} 0 \\ p \end{bmatrix},$$

for a choice of $a_{21}, a_{22} \in \mathbb{R}[s]$. From

$$\det(M) = (s-a)\,a_{22} + b a_{21} = c \in \mathbb{R} - \{0\}$$

one derives the solution

$$\begin{bmatrix} x \\ u \end{bmatrix} = \begin{bmatrix} 0 & b \\ -1/b & (s-a) \end{bmatrix} \begin{bmatrix} 0 \\ p \end{bmatrix}$$

for the choice $c = 1$, which implies $a_{21} = 1/b$, $a_{22} = 0$. Provided $b \neq 0$ is met, we can characterize all system variables again by p and its first time derivative.

The property of flatness allows us also to simplify the controller design. We rewrite (4) in the coordinates y, y_1 and get the simple integrator

$$\dot{y} = y_1 \,. \tag{6}$$

This observation leads to the notion of *state feedback equivalence* of two systems, here the systems (4) and (6) are connected by the invertible map (5), where the inputs are involved. Obviously, trajectory planning for (6) is a trivial task, and any choice $y = y_d(t)$ is admissible, whenever $y_d(t)$ is a differentiable function. With help of (5) one derives the data for (4). It is also straightforward to stabilize this trajectory. Therefore, we introduce the error $e = y - y_d(t)$ and require that $\dot{e} + \alpha e = 0, \alpha > 0$ is met. A simple calculation shows that the feedback $y_1 = \partial_t y_d(t) - \alpha(y - y_d(t))$ for (6) or equivalently $u = -(a/b)\,x + \partial_t y_d(t) - \alpha(x/b - y_d(t))$ for (4) stabilizes the planned trajectory. These design approaches will be more deepened in the following sections.

3 Control loop design

Let us consider a system of the type

$$\dot{x} = f(x, u) \tag{7}$$

with n-dimensional state x and 1-dimensional input u, which is state feedback equivalent to a chain of n integrators. Respectively, there exists a one to one map φ,

$$\begin{aligned} x &= \varphi_x(y, y_1, \ldots, y_{n-1}) \\ u &= \varphi_u(y, y_1, \ldots, y_n) \end{aligned} \tag{8}$$

with coordinate y for the flat output and y_k, $k = 1, \ldots, n$ for the k-times time derivatives such that (7) takes the simple form

$$\begin{aligned} \dot{y} &= y_1 \\ &\vdots \\ \dot{y}_{n-1} &= y_n \end{aligned} \tag{9}$$

rewritten in the coordinates $y, y_1, \ldots, y_n$. Obviously, trajectory planning for (9) is a much easier task than for (7). E.g. let us design a trajectory which joins the initial point x_0 with a terminal point x_T. This is an under determined two point boundary problem, which is hard to solve in general. But exploiting the structure of (9), we choose a polynomial

$$y_d(t) = \sum\nolimits_{i=0}^{2n-1} a_i t^i,$$

which meets the interpolation conditions $\partial_t^i y_d(t) = y_i(t)$ with $x(t) = \varphi_x(y(t), \ldots, y_{n-1}(t)), t \in \{0, T\}$ and we are done. To stabilize this trajectory we introduce the error $e = y - y_d(t)$ together with its derivatives $e_i, i = 1, \ldots, n-1$ and design a control law for (9) such that

$$\begin{aligned} \dot{e} &= e_1 \\ &\vdots \\ \dot{e}_{n-1} &= \sum_{i=0}^{n-1} -\alpha_i e_i \end{aligned}$$

is met with a Hurwitz polynomial $\sum_{i=0}^{n} \alpha_i s^i$, $\alpha_n = 1$.Then one derives the control law for (7) with help of (8).

Let us discuss a more challenging problem now. Like above we construct a trajectory, which joins x_0 with x_T, but additionally we minimize T under the constraint $|u(t)| \leq u_{max}$. We solve the problem in two steps. Let us suppose, we are able to solve the optimization problem

$$\begin{aligned} &\min_{y_d} \ \varepsilon(T) \\ &\quad |\varphi_u(y(t), \ldots, y_n(t))| \leq u_{max}\varepsilon, \ t \in [0, T] \\ &\quad x(t) = \varphi_x(y(t), \ldots, y_{n-1}(t)), \ t \in \{0, T\} \end{aligned} \tag{10}$$

for an on $[0, T]$ n-times differentiable function y_d. Since the terminal time T is considered as a given parameter, we derive the monotonically decreasing function $\varepsilon(T)$, if we vary T. In the second step, we solve the equation $\varepsilon(T) = 1$. To make the first step numerically manageable, we parameterize the function y_d by a family of suitable functions, e.g. polynomials, to derive a relation of the type

$$y_d(t) = y_d(t, p), \quad p \in \mathbb{R}^k$$

where the k parameters p_i determine y_d. Roughly speaking, we convert the infinite dimensional optimization problem to a finite one. Of course solutions of the latter

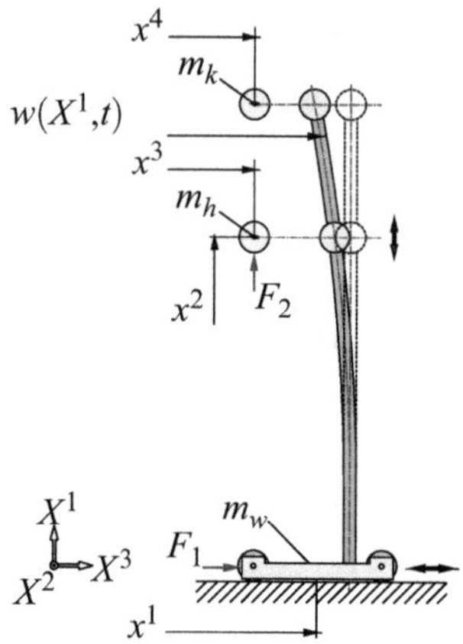

Fig. 1 Single mast stacker crane with coordinates for the model

one are only suboptimal solutions of the first one. But one can show, that the difference between sub- and optimal solution can be made arbitrary small. The following section presents an example with B-splines.

4 Time optimal trajectories for a single mast stacker crane

Let us consider the single mast stacker crane of a high bay warehouse, which is presented in Fig. 1. The control objective for this flexible mechanical structure is to guide the lifting unit from a given starting position to a desired terminal position, where the movement takes place in the $\left(X^1, X^3\right)$-plane and both points are equilibria. This transfer should be as fast as possible, but several constraints must be met because of security and protection of the structure, see [6] also for more details about the following. To derive a mathematical model simple enough for the controller design but sufficiently accurate to cover phenomena like spill over, we make the following assumptions. The tip mass, the lifting and driving unit are modeled as point masses with mass m_k, m_h, m_w. The flexible structure of length l is modeled according to Euler-Bernoulli hypothesis, we assume uniform mass density ρ, cross section surface A, as well as flexural rigidity EI. The rotational inertia is neglected and the geometric relations are linearized. The center of gravity of m_h moves along the beam center line. The control inputs are the forces F_1, F_2, which act on the driving and lifting unit, respectively.

We use the coordinates x^1, x^2, x^3, x^4, w as shown in Fig. 1, and denote the time derivative coordinates by $\dot{x}^i, \ddot{x}^i$, $i = 1, \ldots, 4$, $\dot{w}, \ddot{w}$ and the spatial ones by $w_x, w_{xx}, w_{xxx}, w_{xxxx}$ in a standard manner. The derivation of the equations of motion uses the Lagrangian formalism with the kinetic energy

$$2E_{kin} = \left(m_w(\dot{x}^1)^2 + m_k(\dot{x}^4)^2 + m_h((\dot{x}^3)^2 + (\dot{x}^2)^2)\right) + \int_0^l \dot{w}^2\left(\cdot, X^1\right)\rho A \mathrm{d}X^1,$$

the potential and stored elastic energy

$$E_{pot} = m_h g x^2 + \frac{1}{2}\int_0^l EI w_{xx}^2\left(\cdot, X^1\right)\mathrm{d}X^1.$$

The contribution of the external forces is taken into account by the term

$$E_{ext} = x^1 F_1 + x^2 F_2 .$$

Hence, the Lagrangian reads as $\tilde{L} = E_{kin} - E_{pot} + E_{ext}$,

$$\tilde{L} = \bar{L}(x, \dot{x}) + \int_0^l l\left(w\left(\cdot, X^1\right), \dot{w}\left(\cdot, X^1\right)\right)\mathrm{d}X^1 .$$

To include the restrictions $\Theta^1 = x^3 - w(\cdot, x^2)$, $\Theta^2 = x^4 - w(\cdot, l) = 0$, we apply the Lagrange multiplier techniques and derive the total Lagrangian L,

$$L = \tilde{L} + \lambda_1\Theta^1 + \lambda_2\Theta^2$$

with multipliers λ_1, λ_2. Finally, we end up with the partial differential equation

$$\rho A\ddot{w} + EI w_{xxxx} = 0,$$

the ordinary differential equations

$$\begin{aligned} m_w\ddot{x}^1 + EI w_{xxx}(\cdot, 0) &= F_1 \\ m_h\ddot{x}^2 + m_h g + m_h\ddot{x}^3 w_x(\cdot, x^2) &= F_2 \\ EI(w_{xxx}(\cdot, x^2_-) - w_{xxx}(\cdot, x^2_+)) &= m_h\ddot{x}^3 \\ EI w_{xxx}(\cdot, l) &= m_k\ddot{x}^4, \end{aligned}$$

and the relations

$$w(\cdot, 0) = x^1,\ w_x(\cdot, 0) = 0,\ EI(w_{xx}(\cdot, x^2_-) - w_{xx}(\cdot, x^2_+)) = 0,\ EI w_{xx}(\cdot, l) = 0.$$

Here the right and left-hand limit of x^2 are denoted by x^2_+ and x^2_-.

We use the simple Ritz ansatz

$$w(X^1) = x^1 + \Phi_1(X^1)\bar{q}^1$$

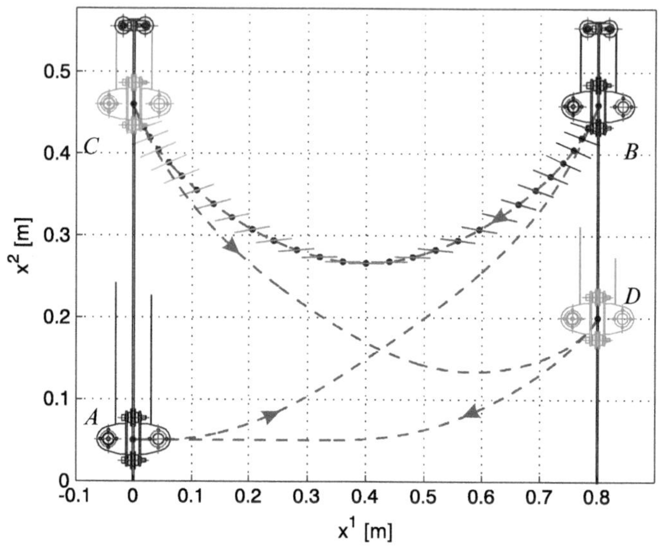

Fig. 2 Some optimal trajectories

$$\Phi(X^1) = 6\left(X^1/l\right)^2 - 3\left(X^1/l\right)^3 + \frac{1}{2}\left(X^1/l\right)^4$$

with the new coordinate $\bar{q}^1$ for the displacement w. Although the shear force conditions at $X^1 = x^2$, $X^1 = l$ are not met, one can show that this simplification is admissible for the feed forward design. The simplified equations of motion follows as

$$M(q)\ddot{q} + C(q,\dot{q})\dot{q} + g = Gu$$

with $q = \left[x^1, \bar{q}^1, x^2\right]^T$ and $u = [F_1, F_2]^T$. In [6] it is shown that this simplified model is flat, and we choose the finite dimensional parametrization

$$y_f^1(t,p) = \sum\nolimits_{i_1=1}^{n_1} B_{i_1}^1(t)p_1^{i_1}, \quad y_f^2(t,p) = \sum\nolimits_{i_2=1}^{n_2} B_{i_2}^2(t)p_2^{i_2}$$

for the flat outputs (y_f^1, y_f^2) by B-splines with parameters $p = \left(p_1^1, \ldots, p_{n_1}^1, p_1^2, \ldots, p_{n_2}^2\right)$, see [6] and the references therein, where $B_j^i(t)$, $i = 1,2$ denotes the B-spline basis Function of the j-th knot with order m_i, such that $y_f^i(t)$ is m_i-times differentiable. Any trajectory of our system must meet constraints $F_{i,\max}, \dot{x}_{\max}^i, \ddot{x}_{\max}^i, \dddot{x}_{\max}^i$, $i = 1,2$ concerning the inputs, the velocities, the accelerations and jerks. Additionally, we have to constrain the bending moment $|M_b| \leq M_{b,\max}$ at the clamping area, where the highest bending stress occurs. Now, we are able to formulate the optimization problem

$$\begin{aligned}&\min_p \varepsilon(T)\\&\quad \left|\dot{x}^i(t,p)\right| \leq \dot{x}^i_{\max}\varepsilon, \left|\ddot{x}^i(t,p)\right| \leq \ddot{x}^i_{\max}\varepsilon, \left|\dddot{x}^i(t,p)\right| \leq \dddot{x}^i_{\max}\varepsilon,\\&\quad \left|F_i(t,p)\right| \leq F_{i,\max}\varepsilon, \left|M_b(,p)\right| \leq M_{\max}\varepsilon,\end{aligned}$$

$i = 1, 2$, $t \in [0, T]$, which is the discretized counterpart to (10). A further simplification is to check the inequalities on grid $t \in g = (0, t_1, \ldots, T)$ only.

Some time optimal trajectories are presented in Fig. 2. They were calculated by a C++ program, which uses a SQP solver of the NAG library, see [6] and the references therein. Further data are $\dim p = 109$, $\dim g = 2001$. It takes less than 0.5 s to determine a time optimal trajectory on a standard desktop computer. Finally, it is worth mentioning that the design of the feedback part of the controller uses the non simplified set of equations. Therefore, problems like spill over, etc. are avoided, see also [6].

References

1. Fliess, M., Levine, J., Martin, P., Rouchon, P.: Flatness and defect of nonlinear systems: introductory theory and examples. Int. J. Contr. **61**, 1327–1361 (1995)
2. Frankel, T.: The Geometry of Physics, 2nd edn. Cambridge University Press, Cambridge (2004)
3. Levine, J.: Analysis and Control of Nonlinear Systems: A Flatness-based Approach. Springer, Berlin (2009)
4. Rudolph, J.: Flatness Based Control of Distributed Parameter Systems. Shaker Verlag, Aachen (2003)
5. Schöberl, M., Rieger, K., Schlacher, K.: System parametrization using affine derivative systems. In: Proceedings 19th International Symposium on Mathematical Theory of Networks & Systems (MTNS), pp. 1737–1743 (2010)
6. Staudecker, M., Schlacher, K., Hansl, R.: Passivity based control and time optimal trajectory planning of a single mast stacker crane. In: Proceedings of the 17th IFAC World Congress, 2008. DOI: 10.3182/20080706-5-KR-1001.00150

Wave Radiation from a Stick-Slip-Like Source Motion (SH-Wave)

Kazumi Watanabe

Abstract A simple mathematical model for the stick-slip motion is introduced and the dynamic response of an elastic solid to a moving SH-source with the stick-slip motion is discussed. Its exact response is obtained in the form of an infinite sum of time-harmonic response. Each component in the summation has a parametric-like frequency, i.e. a combination of two frequencies, stick-slip motion and time-harmonic source magnitude. The Doppler frequency shift is extracted from each component response. Due to the presence of two velocity parameters, the mean velocity and the frequency of the stick-slip motion, the Doppler effects take place even if the source magnitude is constant, not time-harmonic. The response is shown in graphical forms and it is found that the lowest frequency component has the dominant contribution to the total response.

1 Introduction

Dynamic response to a moving load or source is one of attractive research fields due to its wide application in engineering and seismology. The work for a moving load on/in an elastic solid dates back to that of Cole and Huth [1]. Since then, many works have been carried out and its source motion has also been extended to non-uniform motion, such as back & fourth, acceleration and deceleration.

Further, if we consider the slip on a non-flat zigzag face, a stick-slip motion is expected as the wave source. Elastodynamic problem concerning the source with stick-slip motion is very scarcely. Since the dynamic response to the stick-slip

K. Watanabe (✉)
Department of Mechanical Engineering, Yamagata University,
Yonezawa, Yamagata 992-8510, Japan
e-mail: kazy@yz.yamagata-u.ac.jp

H. Irschik et al. (eds.), *Advanced Dynamics and Model-Based Control of Structures and Machines*, DOI 10.1007/978-3-7091-0797-3_24,

source is more realistic and useful, more attention should be directed to the elastodynamic problem of the moving source with stick-slip motion. This is the first motivation for the present work.

It has been well-known that a running time-harmonic source [2] or slip [3] produces the Doppler effects in the response, since two velocity parameters, frequency and moving velocity, exist in the problem setting. If we consider a source with the stick-slip motion, two velocity parameters also exist. They are the mean velocity and the frequency of the stick-slip motion and the Doppler effects are expected in the response. Thus, it is interesting to find how the stick-slip motion produces the Doppler effects. This is the second motivation of the work.

2 Modeling for stick-slip motion

A typical velocity variation for the stick-slip motion is shown in Fig. 1a [4]. It has a finite velocity gradient at the starting point of the slip, but does zero gradient at the end of the slip. In the sticking state, the velocity vanishes. After a short sticking, the slip and stick repeat the same velocity variation. It will be not easy to represent the actual velocity variation with simple mathematical formulas. However, we recall the simple trigonometric function,

$$v(t) = V\{1 - \cos\omega_2 t\}, \tag{1}$$

and its variation is very close to that of the actual stick-slip motion shown in Fig. 1b. This function cannot represent the initial velocity gradient of the slip, but does the zero gradient at the end of the slip. Further, this formula can represent the frequency of stick-slip motion with ω_2 and the mean velocity with V. Thus, the simple formula (1) for the stick-slip motion is adopted here as the first step for the modeling.

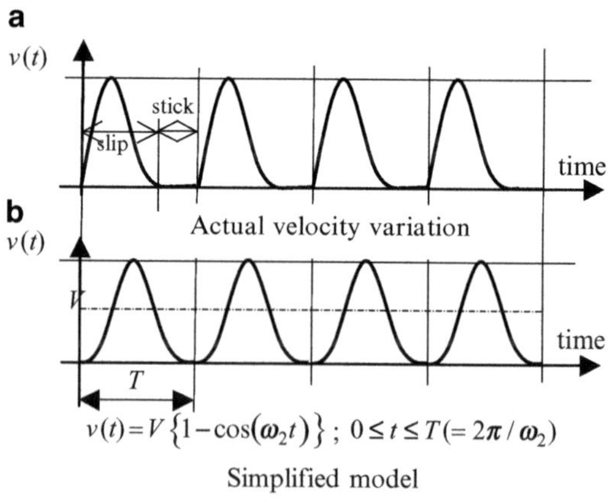

Fig. 1 Actual and model velocities of the stick-slip motion

The location of load/source is obtained by the integration,

$$L(t) = \int v(t)dt = \frac{V}{\omega_2}\{\omega_2 t - \sin\omega_2 t\}, \tag{2}$$

where $L(0)$ at $t = 0$ is a reference position. When a body force is moving on the x-axis in an infinite elastic solid and shows the stick-slip motion, the location of the force is expressed by the Diracs delta function $\delta(x - L(t))$ where $L(t)$ is defined by (2). Further, we assume that the magnitude of the body force is time-harmonic with frequency ω_1. Then, the body force directed to the j-axis is given by the formula,

$$B_j \exp(+i\omega_1 t)\delta\left(x - \frac{V}{\omega_2}\{\omega_2 t - \sin\omega_2 t\}\right)\delta(y), \quad j = x, y, z. \tag{3}$$

This is our mathematical model for the stick-slip motion of the body force.

3 Response to a moving source with stick-slip motion

Let us consider an anti-plane deformation of an infinite elastic solid. We assume that a moving body force on the x-axis shows the stick-slip motion and is time-harmonic. Then the body force is expressed by (3) and SH-wave, anti-plane displacement u_z, is governed by a particular solution of the non-homogeneous wave equation,

$$\begin{aligned}\frac{\partial^2 u_z}{\partial x^2} + \frac{\partial^2 u_z}{\partial y^2} = \frac{1}{c_s^2}\frac{\partial^2 u_z}{\partial t^2}\\ -\frac{Q}{c_s^2}\exp(+i\omega_1 t)\delta\left(x - \frac{V}{\omega_2}\{\omega_2 t - \sin\omega_2 t\}\right)\delta(y).\end{aligned} \tag{4}$$

In order to obtain its particular solution, we apply triple Fourier transform with respect to all variables (x, y, t) defined by

$$\begin{aligned}
\tilde{f}(\xi) &= \int_{-\infty}^{+\infty} f(x)\exp(+i\xi x)dx \Leftrightarrow f(x) = \frac{1}{2\pi}\int_{-\infty}^{+\infty} \tilde{f}(\xi)\exp(-i\xi x)d\xi\\
\bar{f}(\eta) &= \int_{-\infty}^{+\infty} f(y)\exp(+i\eta y)dy \Leftrightarrow f(y) = \frac{1}{2\pi}\int_{-\infty}^{+\infty} \bar{f}(\eta)\exp(-i\eta x)d\eta\\
f^*(\lambda) &= \int_{-\infty}^{+\infty} f(t)\exp(+i\lambda t)dt \Leftrightarrow f(t) = \frac{1}{2\pi}\int_{-\infty}^{+\infty} f^*(\lambda)\exp(-i\lambda t)d\lambda
\end{aligned} \tag{5}$$

and then make the use of the formula for the Bessel function [5],

$$J_n(z) = \frac{1}{2\pi}\int_0^{2\pi} \exp(+in\theta - iz\sin\theta)d\theta, \tag{6}$$

the transformed displacement is given by the sum of definite integrals,

$$\begin{aligned}\bar{\tilde{u}}_z^* &= \frac{Q/c_s^2}{\xi^2+\eta^2+(\lambda/c_s)^2}\\ &\times \sum_{n=-\infty}^{+\infty} \delta(\lambda+\omega_1+V\xi-n\omega_2)\int_0^{2\pi} \exp\{+in\varphi - i(\xi V/\omega_2)\sin\varphi\}d\varphi. \end{aligned}\tag{7}$$

Our next task is to perform the inversion of the triple Fourier transform. The triple Fourier inversion is carried out successively. The first Fourier inversion with respect to variable η, is performed by applying the formula [6],

$$\frac{1}{2\pi}\int_{-\infty}^{+\infty} \frac{1}{x^2+a^2}\exp(-ibx)dx = \frac{1}{2a}\exp(-a|b|), \tag{8}$$

and it yields to

$$\begin{aligned}\tilde{u}_z^* &= \frac{Q/c_s^2}{2q}\exp(-q|y|)\\ &\times \sum_{n=-\infty}^{+\infty} \delta(\lambda+\omega_1+V\xi-n\omega_2)\int_0^{2\pi} \exp\{+in\varphi - i(\xi V/\omega_2)\sin\varphi\}d\varphi, \end{aligned}\tag{9}$$

where

$$q = \sqrt{\xi^2-(\lambda/c_s)^2}\,, \quad \mathrm{Re}(q) \geqslant 0. \tag{10}$$

The second inversion integral with respect to λ is also easily carried out with aids of the integral formula for the Delta function,

$$\int_a^b f(\lambda)\delta(\lambda-c)d\lambda = \begin{cases} f(c)\,; & a<c<b \\ 0\,; & c<a \quad \text{or} \quad b<c\,. \end{cases} \tag{11}$$

Then, we have

$$\tilde{u}_z = \frac{Q}{4\pi c_s^2} \sum_{n=-\infty}^{+\infty} \exp\{-i(n\omega_2 - \omega_1)t\} \int_0^{2\pi} \exp(+in\varphi)d\varphi$$
$$\times \frac{\exp\left[-|y|\sqrt{\xi^2 - \left\{\frac{n\omega_2-\omega_1-V\xi}{c_s}\right\}^2} + i\xi\left\{Vt - \frac{V}{\omega_2}) \sin\varphi\right\}\right]}{\sqrt{\xi^2 - \left\{\frac{n\omega_2-\omega_1-V\xi}{c_s}\right\}^2}}. \quad (12)$$

The formal form of the last Fourier inversion with respect to ξ is given by

$$u_z = \frac{Q}{4\pi c_s^2} \sum_{n=-\infty}^{+\infty} \exp\{-i(n\omega_2 - \omega_1)t\} \int_0^{2\pi} \exp(+in\varphi)d\varphi$$
$$\times \frac{1}{2\pi} \int_{-\infty}^{+\infty} \frac{\exp\left[-|y|\sqrt{\xi^2 - \left\{\frac{n\omega_2-\omega_1-V\xi}{c_s}\right\}^2} + i\xi\{Vt - \frac{V}{\omega_2}\sin\varphi - x\}\right]}{\sqrt{\xi^2 - \left\{\frac{n\omega_2-\omega_1-V\xi}{c_s}\right\}^2}} d\xi. \quad (13)$$

In order to discuss the wave nature, the inner integral with respect to the variable ξ must be evaluated. Introducing the simple variable change, $\xi \to \zeta$,

$$\varsigma = \xi + (M/m^2)(n\omega_2 - \omega_1)/c_s\,, \quad m = \sqrt{1 - M^2} \quad (14)$$

where $M\ (= V/c_s)$ is Mach number, and applying the integration formula [2],

$$\exp(+i\omega t)\frac{1}{2\pi}\int_{-\infty}^{+\infty} \frac{1}{\sqrt{\varsigma^2 - (\omega/c)^2}} \exp\left\{-y\sqrt{\varsigma^2 - (\omega/c)^2} - i\varsigma x\right\} d\varsigma$$
$$= \exp(+i\omega t) \begin{cases} -\frac{i}{2}H_0^{(2)}\left(\frac{\omega}{c}\sqrt{x^2+y^2}\right)\ ; & \omega > 0 \\ +\frac{i}{2}H_0^{(1)}\left(\frac{|\omega|}{c}\sqrt{x^2+y^2}\right)\ ; & \omega < 0 \\ -\frac{1}{2\pi}\log\left(x^2+y^2\right)\ ; & \omega = 0, \end{cases} \quad (15)$$

where $H_0^{(1/2)}(.)$ is the zero-th order Hankel function of the 1st or 2nd kind. Thus, the inner integral in (13) is evaluated and the final form is given by

$$u_z = \frac{Q}{8\pi c_s^2} \sum_{n=-\infty}^{+\infty} \exp\left\{-i\frac{n\omega_2 - \omega_1}{m^2}(t - Mx/c_s)\right\}$$

$$\times \int_0^{2\pi} \exp\{+in\varphi + i(M/m)^2(n - \omega_1/\omega_2)\sin\varphi\} d\varphi$$

$$\times \begin{cases} +\dfrac{i}{m} H_0^{(1)}\left(\dfrac{n\omega_2 - \omega_1}{c_s m^2}\sqrt{\{x - Vt + (V/\omega_2)\sin\varphi\}^2 + (my)^2}\right); \; n\omega_2 - \omega_1 > 0 \\ -\dfrac{i}{m} H_0^{(2)}\left(\dfrac{\omega_1 - n\omega_2}{c_s m^2}\sqrt{\{x - Vt + (V/\omega_2)\sin\varphi\}^2 + (my)^2}\right); \; n\omega_2 - \omega_1 < 0 \\ -\dfrac{1}{\pi m}\log\left(\{x - Vt + (V/\omega_2)\sin\varphi\}^2 + (my)^2\right); \; n\omega_2 - \omega_1 = 0. \end{cases} \tag{16}$$

4 Doppler effects

In (16), the variable $x - Vt$ implies the simple coordinate shift along the x-axis and the term with $n\omega_2 - \omega_1 = 0$ shows the static deformation, no wave nature. Then, the term with non-vanishing condition $n\omega_2 - \omega_1 \neq 0$ is useful for our discussion on the Doppler effects, but has somewhat complicated form. In order to discuss the Doppler effects, an asymptotic form for this dynamic part of the response is derived.

Assuming that the observing point is sufficiently apart from the moving source, the Hankel functions are approximated as

$$\left.\begin{array}{l} H_0^{(1)}\left(\frac{n\omega_2 - \omega_1}{c_s m^2}\sqrt{\{x - Vt + (V/\omega_2)\sin\varphi\}^2 + (my)^2}\right) \\ H_0^{(2)}\left(\frac{\omega_1 - n\omega_2}{c_s m^2}\sqrt{\{x - Vt + (V/\omega_2)\sin\varphi\}^2 + (my)^2}\right) \end{array}\right\}$$

$$\approx \sqrt{\frac{2c_s m^2}{\pi r|n\omega_2 - \omega_1|}} \exp\left(+i\frac{n\omega_2 - \omega_1}{c_s m^2} r\right), \tag{17}$$

where $r = \sqrt{(x - Vt)^2 + (my)^2}$. Then, the asymptotic form of the dynamic response is given by

$$u_z \sim \frac{iQ}{8\pi m c_s^2} \sum_{\substack{n=-\infty \\ n\omega_2 - \omega_1 \neq 0}}^{+\infty} \exp\left[-i\underline{\underline{\frac{n\omega_2 - \omega_1}{m^2}\{t - M(x/c_s) - r/c_s\}}}\right] \operatorname{sgn}(n\omega_2 - \omega_1)$$

$$\times\sqrt{\frac{2c_s m^2}{\pi r|n\omega_2-\omega_1|}}\int_0^{2\pi}\exp\{+in\varphi+i(M/m)^2(n-\omega_1/\omega_2)\sin\varphi\}d\varphi. \tag{18}$$

In the above equation, the integration term is a simple coefficient and does not represent any wave nature. The phase function, underlined exponent of the exponential function,

$$\Theta(t)=\frac{n\omega_2-\omega_1}{m^2}\left\{t-Mx/c_s-\sqrt{(x-Vt)^2+(my)^2}/c_s\right\}, \tag{19}$$

represents the wave nature. Since the time variable t is included in the square root, we define the "instantaneous" frequency $\omega(t)$ as

$$\omega(t)=\frac{\partial\Theta(t)}{\partial t}=\frac{n\omega_2-\omega_1}{1-M^2}\left\{1+M\frac{x-Vt}{\sqrt{(x-Vt)^2+(my)^2}}\right\}. \tag{20}$$

The instantaneous frequency dependents on the location of the observing point, $x-Vt$ and y. If an observer is on the line of the source trajectory, the frequencies before and after the source are given by

$$\omega(t)=\begin{cases}\frac{n\omega_2-\omega_1}{1-M}; & x-Vt>0,\ y=0\\ \frac{n\omega_2-\omega_1}{1+M}; & x-Vt<0,\ y=0\end{cases};\quad n=0,\pm1,\pm2,... \tag{21}$$

This is the typical Doppler frequency shift from its reference frequency $n\omega_2-\omega_1$. This reference frequency is a parametric-like combination of two frequencies, frequency of the time-harmonic source magnitude and that of the stick-slip motion. Since the summation is over all indices n, we have an infinite sequence of the reference frequencies even if the source magnitude is constant ($\omega_1=0$). This states that the stick-slip motion of the source produces the Doppler effects and its reference frequency is the parametric-like frequency combination, $n\omega_2-\omega_1$, not the source frequency ω_1 itself. It is very interesting that the Doppler effects take place even if the source magnitude is constant, $\omega_1=0$. This is a characteristic feature of the stick-slip motion.

Figure 2 shows a dynamic response in the case of constant source magnitude. Thin light curves in the figure show the sum of two components with the summation indices $\pm n$, and the thick curve does the total response where the summation is carried our from $n=-10$ to $n=+10$. Since the observing point is on the y-axis, the response before the source shows higher frequency, but that after the source does lower frequency. The response changes its frequency at time $t=0$. This clear frequency discrepancy in the response is the typical Doppler effects.

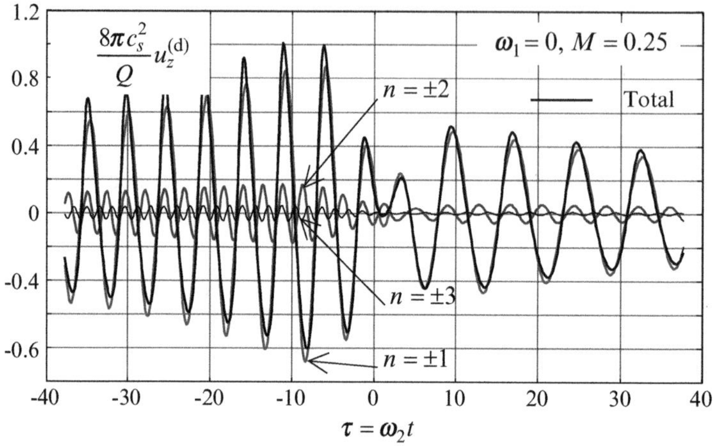

Fig. 2 A typical time response and its components at $(x = 0, \omega_2 y/(2\pi V) = 1)$

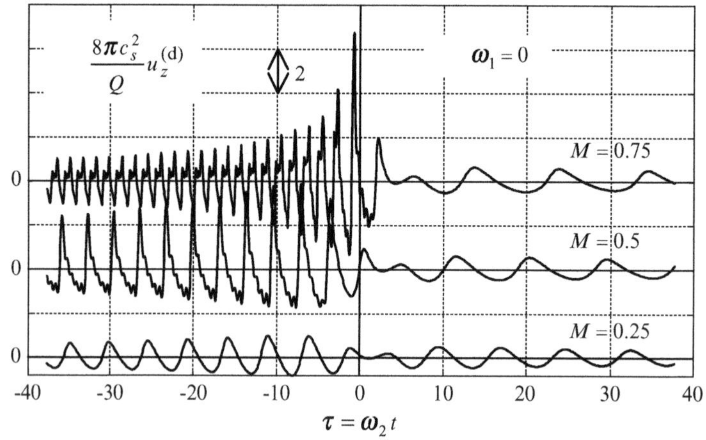

Fig. 3 Time responses at $(x = 0, \omega_2 y/(2\pi V) = 1)$ for three cases of Mach number

Figure 3 shows the dynamic response for three Mach numbers. The response before the source shows more complicated wave form for the larger Mach number, but that after the source does smoother wave form.

5 Conclusion

Introducing a simple mathematical model for the stick-slip motion, SH-wave radiation from a point source with stick-slip motion is discussed. The stick-slip motion of the source produces the Doppler frequency shifts even if the source magnitude is constant, not time harmonic. There exists an infinite sequence of the reference frequency for the Doppler effects. The reference frequency is a parametric-like combination of two frequencies, stick-slip motion and source magnitude.

References

1. Cole, J., Huth, J.: Trans. ASME, J. Appl. Mech. **25**, 433–436 (1958)
2. Watanabe, K., Biwa, S.: Acta Mech. **195**, 27–59 (2008)
3. Ben-Menahem, A., Singh, S.J.: Seismic Waves and Sources, pp. 237, 305–307. Springer, New York (1981)
4. Halling, J. (ed.): Principles of Tribology, pp. 151–152. The MacMillan Press, London (1978)
5. Watson, G.N.: Theory of Bessel Functions, 2nd edn. p. 31. Cambridge University Press, New York (1966)
6. Erdelyi, A. (ed.): Tables of Integral Transforms, Vol. 1, p. 8. McGraw-Hill, New-York (1954)

Derivation of the Instability Potential of Plates by Rigid Body and Equilibrium Considerations

Y.B. Yang, S.R. Kuo, J.L. Peng, and M.H. Shih

Abstract A new method is proposed for deriving the instability potential of initially stressed plates based on the rigid body and equilibrium considerations using the updated Lagrangian formulation. Starting from the *rigid body rule*, the virtual instability potential was derived for a *thin plate* under *real* rigid displacements. Next, utilizing the *equilibrium equations* for the boundary forces at the C_1 and C_2 states, another virtual instability potential was derived for the plate under *virtual* rigid displacements. Comparing the two potentials yields the one in *total form* for the plate. The present approach requires only simple integrations and analogical comparison of related virtual works, thereby avoiding the physically unclear, complicated derivations involved in previous procedures. The present theory was demonstrated to be more reliable for the postbuckling analysis of structures.

1 Introduction

In an incremental nonlinear analysis of solid structures, three typical configurations are considered, i.e., the initial configuration C_0, last configuration C_1, and current configuration C_2. Assuming that the behavior of a structure, including deformations and loadings, from C_0 to C_1 is known, one is interested in its behavior during the

Y.B. Yang (✉) · J.L. Peng
Department of Construction Engineering, National Yunlin University of Science and Technology, Douliu, Yunlin 64002, Taiwan
e-mail: ybyang@ntu.edu.tw

S.R. Kuo
Department of Harbor and River Engineering, National Taiwan Ocean University, Keelung 20224, Taiwan

M.H. Shih
Department of Civil Engineering, National Chi-Nan University, Puli, Nantou 545, Taiwan

H. Irschik et al. (eds.), *Advanced Dynamics and Model-Based Control of Structures and Machines*, DOI 10.1007/978-3-7091-0797-3_25,

incremental step from C_1 to C_2 under a load increment [1]. In this paper, the updated Lagrangian (UL) formulation is adopted to derive the incremental nonlinear theory for plate structures, by which C_1 is selected as the configuration of reference.

One advantage with the virtual work methods for deriving the stability theory of a solid element is that the procedure can be carried out in a systematic and routine manner. Namely, by substituting the strain-displacement relations, stress–strain relations, and force-stress relations for the element into the virtual work equation, the governing differential equations, along with the natural and geometric boundary conditions, can be derived by utilizing the arbitrary nature of variations. However, for the plate element of concern here, the above procedure is not foolproof. There are several issues that must be adequately addressed before a successful result can be obtained. First, when dealing with the virtual work related to the six nonlinear virtual strains, there always appear some terms with unclear physical meanings, as a result of the adoption of different hypotheses for plane sections after deformation. The other issue is concerned with the derivation of equivalent shear forces on the boundaries of the plate at C_2, which may differ for different hypotheses of plate sections. It is realized that only when the boundary shear forces are correctly specified at C_2 can Kirchhoff's forces be accurately computed for the plate element, and that the latter is essential to derivation of the virtual works done by surface tractions on the plate at C_2 [2]. A frequently encountered situation is that even if a nonlinear plate theory was deliberately derived, there is no guarantee that the theory can pass the *rigid body rule* [1, 3], a fundamental property to be obeyed by all initially stressed elements.

The objective of this paper is to present a new, simple method, based on the fundamental principles of rigid body rule and force equilibrium, for deriving the virtual instability potential of a thin plate, under the conditions of either given *real* or *virtual* rigid displacements. By an analogical comparison of the two sets of virtual instability potential derived, the *total* instability potential can be derived for the plate. Due to the lack of similar approaches existing in the literature, no effort will be made herein to review related previous works.

2 Method of formulation

By the UL formulation, the virtual work equation for the plate can be written with reference to C_1 as follows [1]:

$$\int_{^1V} C_{klij} e_{kl} \delta e_{ij} \, dV + \int_{^1V} \tau_{ij} \delta \eta_{ij} \, dV = \oint_{^1S} {}^2t_i \delta u_i \, dS - \oint_{^1S} {}^1t_i \delta u_i \, dS \tag{1}$$

or

$$\delta U_E + \delta U_G = \delta W_2 - \delta W_1 \tag{2}$$

where δU_E = virtual linear strain energy, δU_G = *virtual instability potential* of the plate, δW_2 = virtual work done by the surface tractions ${}^2t_i$ acting at C_2, and δW_1 = virtual work done by the surface tractions ${}^1t_i$ acting at C_1. Other symbols in (1) are defined as follows: δ = variation; C_{klij} = material coefficients, e_{kl} = linear strain components, τ_{ij} = Cauchy stresses; η_{ij} = nonlinear strain components; u_i = displacement increments; ${}^1V$ = volume; and ${}^1S$ = surface area of the element. Specifically, the virtual linear strain energy δU_E is a function of the displacement increments $\tilde{u}$ and virtual displacements $\delta\tilde{u}$ of the plate element, namely,

$$\delta U_E = \Pi_E(\tilde{u}, \delta\tilde{u}); \qquad \tilde{u} = u_x\mathbf{e}_x + u_y\mathbf{e}_y + u_z\mathbf{e}_z \tag{3}$$

Since the internal forces existing at C_1 are known, we can express the virtual instability potential δU_G and external virtual work δW_1 at C_1 as follows:

$$\delta U_G = \Pi_G(\tilde{u}, \delta\tilde{u}) \tag{4}$$

$$\delta W_1 = \Pi_1(\delta\tilde{u}) \tag{5}$$

Meanwhile, since the external force increments at C_2 are related to the displacement increments $\tilde{u}$, we can express the external virtual work δW_2 at C_2 as a function of the displacement increments $\tilde{u}$ and virtual displacements $\delta\tilde{u}$, i.e.,

$$\delta W_2 = \Pi_2(\tilde{u}, \delta\tilde{u}) \tag{6}$$

In this paper, we shall use the incremental virtual work equation in (2), together with the fundamental principles of rigid body rule and equilibrium conditions, to derive the virtual instability potential $\Pi_G(\tilde{u}_r, \delta\tilde{u})$ of the plate for given *real* rigid displacements and the one $\Pi_G(\tilde{u}, \delta\tilde{u}_r)$ for given *virtual* rigid displacements. Then, by analogically comparing the two expressions derived for the plate under the same actions, we can derive the *total* virtual instability potential $\Pi_G(\tilde{u}, \delta\tilde{u})$ of the plate that satisfies both the rigid body rule and force equilibrium conditions.

In the following, we shall illustrate how the following three terms can be derived: the virtual instability potential $\Pi_G(\tilde{u}_r, \delta\tilde{u})$ of the plate for given *real* rigid displacements, the one $\Pi_G(\tilde{u}, \delta\tilde{u}_r)$ for given *virtual* rigid displacements, and the *total* instability potential $\Pi_G(\tilde{u}, \delta\tilde{u})$. Due to restriction of paper length, all the static and kinematic descriptions of the plate (Fig. 1) in the pre-buckling and bucking stages will be omitted.

3 Derivation of $\Pi_G(\tilde{u}, \delta\tilde{u}_r)$ based on rigid body rule

The principle of *objectivity* implies that using different references to describe the behavior of a solid under the same actions, the same result should be obtained. The *rigid body rule* proposed by Yang and Chiou [3] for nonlinear finite elements can be regarded as an example of objectivity, which means that when a finite element

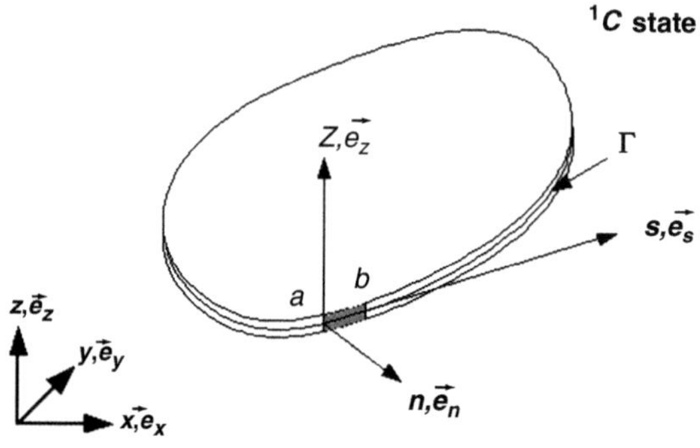

Fig. 1 Thin plate at C_1 and coordinates

equilibrated by a set of forces at C_1 is subjected to a rigid rotation, the forces acting on the element at C_2 should rotate following the rigid rotation, while their magnitudes remain unchanged.

The purpose of this section is to derive from the rigid body rule the *external virtual work* $\delta W_r(u_r, \delta u)$ done by the boundary tractions and the corresponding *virtual instability potential* $\delta \tilde{U}_G = \Pi_G(\tilde{u}_r, \delta \tilde{u})$ for a plate under given *real* rigid displacements. The procedure is as follows. First, we can write down the virtual work δW_1 done by the boundary forces at C_1. For a plate undergoing a rigid displacement, we can also write out the virtual work $\delta \bar{W}_2$ done by the surface tractions after the rigid body rotation at C_2. Subtracting δW_1 from $\delta \bar{W}_2$ yields the *virtual work increment* δW_r done by the boundary forces during the rigid displacement from C_1 to C_2.

For an incremental step containing only rigid displacement, the *linear virtual strain energy* vanishes, i.e., $\delta \bar{U}_E = 0$. Under this condition, we can obtain from (2) the following relation for the *virtual instability potential* $\delta \bar{U}_G$ and the virtual work increment δW_r done by the boundary forces as

$$\begin{aligned} \delta \bar{U}_G &= \Pi_G(\tilde{u}_r, \delta \tilde{u}) = \delta W_r \\ &= \oint ({}^1N_z \theta_{yr} - {}^1N_y \theta_{zr}) \delta u \, \delta s \\ &\quad + \oint ({}^1N_x \theta_{zr} - {}^1N_z \theta_{xr}) \delta v \, \delta s \\ &\quad + \oint ({}^1N_y \theta_{xr} - {}^1N_x \theta_{yr}) \delta w \, \delta s \end{aligned} \tag{7}$$

The preceding equation is exactly the *condition* for the virtual instability potential of the plate to satisfy during a rigid displacement.

The virtual instability potential $\Pi_G(\tilde{u}_r, \delta \tilde{u})$ in (7) is a boundary integral for all the forces acting on the boundary of the plate, which can be transformed by Green's theorem into an area integral in terms of the stress resultants and generalized strains existing over the plate. The virtual work terms in (7) are done by the in-plane actions ${}^1N_x$, ${}^1N_y$ and boundary actions ${}^1N_z$ (including the equivalent shears). For the present purposes, we shall divide the virtual work in (7) into two parts related to the *in-plane* and *out-of-plane actions* as

$$\delta\bar{U}_G = \left(\delta\bar{U}_G\right)_I + \left(\delta\bar{U}_G\right)_O \tag{8}$$

where after some lengthy operations, the two virtual work terms on the right side can be derived as follows:

$$\begin{aligned}
\left(\delta\bar{U}_G\right)_I = {} & \theta_{xr}\iint\left({}^1N_{yy}\delta\frac{\partial w}{\partial y} + {}^1N_{xy}\delta\frac{\partial w}{\partial x}\right)dA \\
& -\theta_{yr}\iint\left({}^1N_{xx}\delta\frac{\partial w}{\partial x} + {}^1N_{xy}\delta\frac{\partial w}{\partial y}\right)dA \\
& +\theta_{zr}\iint\left({}^1N_{xx}\delta\frac{\partial v}{\partial x} + {}^1N_{xy}\delta\frac{\partial v}{\partial y}\right)dA \\
& -\theta_{zr}\iint\left({}^1N_{yy}\delta\frac{\partial u}{\partial y} + {}^1N_{xy}\delta\frac{\partial u}{\partial x}\right)dA
\end{aligned} \tag{9}$$

$$\begin{aligned}
\left(\delta\bar{U}_G\right)_O = {} & \iint\left[{}^1M_{xx}\left(\frac{\partial w_r}{\partial y}\delta\frac{\partial^2 v}{\partial x^2} + \frac{\partial w_r}{\partial x}\delta\frac{\partial^2 u}{\partial x^2}\right)\right. \\
& +{}^1M_{yy}\left(\frac{\partial w_r}{\partial y}\delta\frac{\partial^2 v}{\partial y^2} + \frac{\partial w_r}{\partial x}\delta\frac{\partial^2 u}{\partial y^2}\right) \\
& \left.+2\,{}^1M_{xy}\left(\frac{\partial w_r}{\partial y}\delta\frac{\partial^2 v}{\partial x\partial y} + \frac{\partial w_r}{\partial x}\delta\frac{\partial^2 u}{\partial x\partial y}\right)\right]dA \\
& -\oint{}^1M_n\left(\frac{\partial w_r}{\partial y}\delta\frac{\partial v}{\partial n} + \frac{\partial w_r}{\partial x}\delta\frac{\partial u}{\partial n}\right)ds
\end{aligned} \tag{10}$$

4 Derivation of $\Pi_G(\tilde{u}, \delta\tilde{u}_r)$ based on force equilibrium

Equilibrium is a property that must be obeyed by a solid element at any step of an incremental nonlinear analysis. With the UL formulation, the linear virtual strain energy δU_E is independent of the nodal forces at C_1, but the virtual instability potential δU_G is a function of the initial nodal forces at C_1. Specifically, the virtual instability potential δU_G should be equal to the virtual work done by the force increments in connection with the displacement increments during the incremental step from C_1 to C_2. The following is the procedure for deriving $\delta\tilde{U}_G = \Pi_G(\tilde{u}, \delta\tilde{u}_r)$ given the *virtual* rigid displacements.

First, six force equilibrium equations can be written for the plate each at C_1 and C_2. For a virtual displacement field consisting of three virtual rigid displacements $\delta\Delta_{xr}, \delta\Delta_{yr}, \delta\Delta_{zr}$ and three virtual rotations $\delta\Theta_{xr}, \delta\Theta_{yr}, \delta\Theta_{zr}$, one can write down the virtual works $\delta\tilde{W}_1$ and $\delta\tilde{W}_2$ done by the boundary forces at C_1 and C_2. By taking into account the equilibrium conditions at C_1, one finds that the virtual work of the plate $\delta\tilde{W}_1$ at C_1 vanishes, i.e., $\delta\tilde{W}_1 = 0$. Similarly, by taking into account the equilibrium conditions at C_2, one can derive the virtual work $\delta\tilde{W}_2$. For given virtual rigid displacements, the linear virtual strain energy reduces to zero, i.e., $\delta\tilde{U}_E = \Pi_E(\tilde{u}, \delta\tilde{u}_r) = 0$. From (2), one finds that the virtual instability potential $\delta\tilde{U}_G = \Pi_G(\tilde{u}, \delta\tilde{u}_r)$ can be related to the virtual work increment δW_b done by the boundary forces of the plate as

$$\begin{aligned}\Pi_G(\tilde{u}, \delta\tilde{u}_r) &= \delta\tilde{U}_G = \delta W_b = \delta\tilde{W}_2 - \delta\tilde{W}_1 \\ &= \delta\Theta_{xr} \oint (w\,{}^1N_y - v\,{}^1N_z + \Theta_z^* n_x\,{}^1M_n)\,ds \\ &\quad + \delta\Theta_{yr} \oint (-w\,{}^1N_x + u\,{}^1N_z + \Theta_z^* n_y\,{}^1M_n)\,ds \\ &\quad + \delta\Theta_{zr} \oint (v\,{}^1N_x - u\,{}^1N_y - \Theta_n\,{}^1M_n)\,ds \end{aligned} \tag{11}$$

The preceding equation has been expressed as boundary integrals, which can be transformed by Green's theorem into area integrals and divided into two parts related to the in-plane and out-of-plane actions as follows:

$$\delta\tilde{U}_G = \left(\delta\tilde{U}_G\right)_I + \left(\delta\tilde{U}_G\right)_O \tag{12}$$

where, after lengthy operations, the two terms on the right side can be written as

$$\begin{aligned}\left(\delta\tilde{U}_G\right)_I = &\iint \left[{}^1N_{xx}\left(\frac{\partial w}{\partial x}\delta\frac{\partial w_r}{\partial x}\right) + {}^1N_{yy}\left(\frac{\partial w}{\partial y}\delta\frac{\partial w_r}{\partial y}\right)\right. \\ &\left. + {}^1N_{xy}\left(\frac{\partial w}{\partial y}\delta\frac{\partial w_r}{\partial x} + \frac{\partial w}{\partial x}\delta\frac{\partial w_r}{\partial y}\right)\right] dA \\ &+ \iint \left[{}^1N_{xx}\left(\frac{\partial v}{\partial x}\delta\frac{\partial v_r}{\partial x}\right) + {}^1N_{yy}\left(\frac{\partial u}{\partial y}\delta\frac{\partial u_r}{\partial y}\right)\right. \\ &\left. + {}^1N_{xy}\left(\frac{\partial v}{\partial y}\delta\frac{\partial v_r}{\partial x} + \frac{\partial u}{\partial x}\delta\frac{\partial u_r}{\partial y}\right)\right] dA\end{aligned}$$

$$+\iint\left[{}^{1}N_{xx}\bar{C}_{1}\left(\frac{\partial v}{\partial x}\delta(e_{xy})_{r}\right)+{}^{1}N_{yy}\bar{C}_{2}\left(\frac{\partial u}{\partial y}\delta(e_{xy})_{r}\right)\right]dA$$
$$+\iint\left[{}^{1}N_{xy}\bar{C}_{3}\left(\frac{\partial v}{\partial y}\delta(e_{xy})_{r}\right)+{}^{1}N_{xy}\bar{C}_{4}\left(\frac{\partial u}{\partial x}\delta(e_{xy})_{r}\right)\right]dA \tag{13}$$

$$\begin{aligned}(\delta\tilde{U}_{G})_{O}=&\iint\Bigg[{}^{1}M_{xx}\left(\frac{\partial^{2}v}{\partial x^{2}}\delta\frac{\partial w_{r}}{\partial y}+\frac{\partial^{2}u}{\partial x^{2}}\delta\frac{\partial w_{r}}{\partial x}\right)\\&+{}^{1}M_{yy}\left(\frac{\partial^{2}v}{\partial y^{2}}\delta\frac{\partial w_{r}}{\partial y}+\frac{\partial^{2}u}{\partial y^{2}}\delta\frac{\partial w_{r}}{\partial x}\right)\\&+2\,{}^{1}M_{xy}\left(\frac{\partial^{2}v}{\partial x\partial y}\delta\frac{\partial w_{r}}{\partial y}+\frac{\partial^{2}u}{\partial x\partial y}\delta\frac{\partial w_{r}}{\partial x}\right)\Bigg]dA\\&-\oint{}^{1}M_{n}\left(\frac{\partial w}{\partial y}\delta\frac{\partial v_{r}}{\partial n}+\frac{\partial w}{\partial x}\delta\frac{\partial u_{r}}{\partial n}\right)ds\\&-\oint{}^{1}M_{n}\left(\frac{\partial v}{\partial n}\delta\frac{\partial w_{r}}{\partial y}+\frac{\partial u}{\partial n}\delta\frac{\partial w_{r}}{\partial x}\right)ds\\&+\oint{}^{1}M_{n}\left(\Theta_{z}^{*}n_{x}\delta\frac{\partial w_{r}}{\partial y}-\Theta_{z}^{*}n_{y}\delta\frac{\partial w_{r}}{\partial x}\right)ds\end{aligned} \tag{14}$$

5 Total virtual instability potential $\Pi_G(\tilde{u},\delta\tilde{u}) = \delta U_G(\tilde{u},\delta\tilde{u})$

In previous sections, one has derived the virtual instability potential $\Pi_G(\tilde{u}_r,\delta\tilde{u})$ for given *real* rigid displacements $\tilde{u}_r$ and another potential $\Pi_G(\tilde{u},\delta\tilde{u}_r)$ for given *virtual* rigid displacements $\delta\tilde{u}_r$. These two potentials can be regarded as the special cases of the full potential $\Pi_G(\tilde{u},\delta\tilde{u})$ of the plate with the following relations:

$$\Pi_G(\tilde{u},\delta\tilde{u})|_{u=u_r} = \Pi_G(\tilde{u}_r,\delta\tilde{u}) \tag{15}$$
$$\Pi_G(\tilde{u},\delta\tilde{u})|_{\delta u=\delta u_r} = \Pi_G(\tilde{u},\delta\tilde{u}_r) \tag{16}$$

The preceding two equations imply that the total virtual instability potential $\Pi_G(\tilde{u},\delta\tilde{u})$ satisfies both the rigid body rule and force equilibrium requirements. The other message is that by an analogical comparison of the integral terms containing the same actions in $\Pi_G(\tilde{u}_r,\delta\tilde{u})$ and $\Pi_G(\tilde{u},\delta\tilde{u}_r)$, one can find the corresponding terms for the total potential $\Pi_G(u,\delta u)$ that satisfies the rigid body rule and force equilibrium requirements.

After a term-by-term analogical comparison of the terms in $\Pi_G(\tilde{u}_r, \delta\tilde{u})$ and $\Pi_G(\tilde{u}, \delta\tilde{u}_r)$ containing the same actions, one can obtain the corresponding total form for the virtual instability potential $\Pi_G(u, \delta u)$ of the plate, which for convenience can be expressed separately in terms of the in-plane and out-of-plane actions as follows:

$$\begin{aligned}(\delta U_G)_I = \frac{1}{2}\iint \Bigg[& {}^1N_{xx}\delta\left[\underline{\left(\frac{\partial u}{\partial x}\right)^2} + \left(\frac{\partial v}{\partial x}\right)^2 + \left(\frac{\partial w}{\partial x}\right)^2\right] \\ & + {}^1N_{yy}\delta\left[\left(\frac{\partial u}{\partial y}\right)^2 + \underline{\left(\frac{\partial v}{\partial y}\right)^2} + \left(\frac{\partial w}{\partial y}\right)^2\right] \\ & + 2\,{}^1N_{xy}\delta\left(\frac{\partial v}{\partial x}\frac{\partial v}{\partial y} + \frac{\partial u}{\partial y}\frac{\partial u}{\partial x} + \frac{\partial w}{\partial y}\frac{\partial w}{\partial x}\right)\Bigg] dA \end{aligned} \tag{17}$$

$$\begin{aligned}(\delta U_G)_O = \iint \Bigg[& {}^1M_{xx}\delta\left(\frac{\partial^2 v}{\partial x^2}\frac{\partial w}{\partial y} + \frac{\partial^2 u}{\partial x^2}\frac{\partial w}{\partial x}\right) \\ & + {}^1M_{yy}\delta\left(\frac{\partial^2 v}{\partial y^2}\frac{\partial w}{\partial y} + \frac{\partial^2 u}{\partial y^2}\frac{\partial w}{\partial x}\right) \\ & + 2\,{}^1M_{xy}\delta\left(\frac{\partial^2 v}{\partial x \partial y}\frac{\partial w}{\partial y} + \frac{\partial^2 u}{\partial x \partial y}\frac{\partial w}{\partial x}\right)\Bigg] dA \\ & - \oint {}^1M_n\delta\left(\frac{\partial v}{\partial n}\frac{\partial w}{\partial y} + \frac{\partial u}{\partial n}\frac{\partial w}{\partial x}\right) ds + \oint {}^1M_n(\Theta_z^*\delta\Theta_n)\, ds \end{aligned} \tag{18}$$

The two underlined terms in (17) are included to ensure satisfaction of the tensor property, which have nothing to do with the rigid body or force equilibrium considerations. The last two integrals on the right side of (18) are the two terms that are *new* compared with the existing theories for plates. Because of the presence of these two terms, the virtual instability potential derived, as represented by (17) and (18), is featured by the fact that it is *rigid body-qualified* and conforms to all *equilibrium conditions*.

6 Applications of the present theory

The virtual instability potential of the plate derived herein plus the conventional virtual strain energy, can be used to deal with the rigid body rotational behavior of the plates in the postbuckling range. By the UL-based finite element procedure,

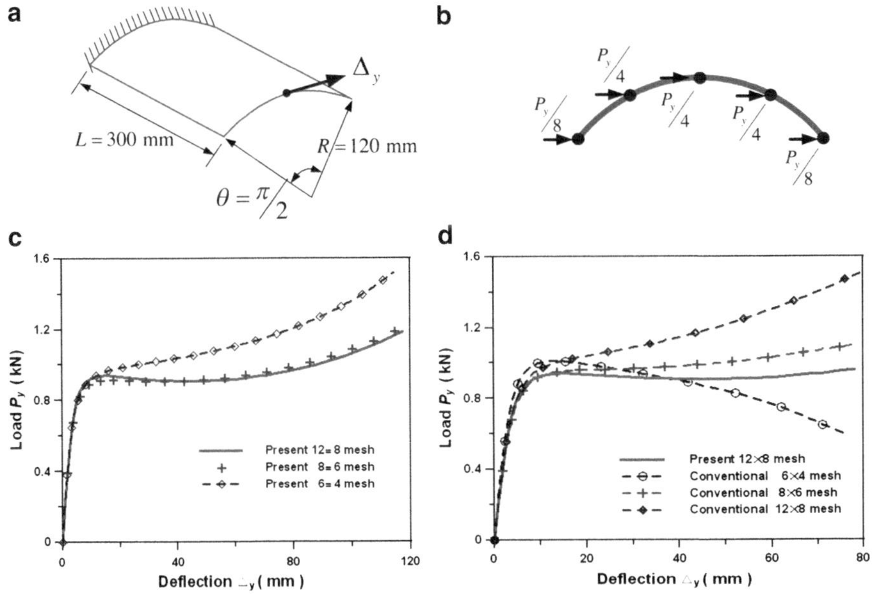

Fig. 2 Cylindrical shell: (**a**) geometry, (**b**) loading, (**c**) mesh test, (**d**) comparison

a *linear stiffness matrix* $[k_e]$ can be derived from the virtual strain energy, and a *geometric stiffness matrix* $[k_g]$ from the virtual instability potential for the plate. The two matrices can be used in combination with a reliable path-tracing scheme, e. g., the *generalized displacement control method* proposed by Yang and Shieh [4], to solve the postbuckling response of plate structures.

Figure 2 shows a cylindrical shell fixed at one end and subjected to horizontal shears at the free end, with the following material properties: elastic modulus $E = 3.10275$ kN/mm^2, thickness $t = 5$ mm, and Poisson's ratio $\nu = 0.3$. Figure 2c shows the convergence characteristics of analysis upon mesh refinement using the present theory. As indicated in Fig. 2d, using the conventional approach, i.e., neglecting the two last terms in (18), does not yield a convergent solution upon mesh refinement. This is mainly due to the fact that the conventional approach fails to cope with rigid rotations in the postbuckling range.

7 Conclusions

A new approach is presented for deriving the virtual instability potential of plates based on the fundamental principles of rigid body rule and force equilibrium using the updated Lagrangian formulation. Compared with the conventional procedures for formulating the plate theory, no assumption on plane sections was needed in the present study. The present theory is superior to the conventional ones in that it can adequately cope with the rigid rotations of shells in the postbuckling range.

Acknowledgements The research reported herein was sponsored largely by the National Science Council through the Taiwan-Austria Joint Research Project with Grant No. 98-2923-E-002-005-MY3.

References

1. Yang, Y.B., Kuo, S.R.: Theory and Analysis of Nonlinear Framed Structures. Prentice-Hall, Singapore (1994)
2. Kuo, S.R., Chi, C.C., Yang, Y.B.: A complete stability theory for Kirchhoff thin plate under all kinds of actions. J. Marine Sci. Tech. **17**(3), 80–193 (2009)
3. Yang, Y.B., Chiou, H.T.: Rigid body motion test for nonlinear analysis with beam elements. J. Eng. Mech., ASCE, **113**(9), 1404–1419 (1987)
4. Yang, Y.B., Shieh, M.S.: Solution method for nonlinear problems with multiple critical points. AIAA J. **28**(12), 2110–2116 (1990)

A Novel Base Isolation System for Asymmetric Buildings in Seismic Active Zones: Damping Supplied by Tuned Liquid Column-Gas Dampers

Franz Ziegler and Basit Khalid

Abstract The vulnerability of civil engineering structures with fundamental frequency, say roughly above 1 Hz, (or buildings with less than 10 stories), when exposed to the strong motion phase of an earthquake is considerably reduced by means of base isolation. The low pass filter for isolating horizontal vibrations is redesigned where the classical elastomeric bearings are substituted by a number of prestressed helical steel springs with pivoted columns along vertical axes carrying part of the dead weight and guiding the horizontal motion. The base isolated building in its fundamental modes is considered to be rigid. Low cost Tuned Liquid Column Gas Dampers, in optimal arrangement in the plan of the basement of the building, supply the effective damping of the remaining horizontal vibrations. The action of the passive damping device is commonly considered to be sufficient. Since the gas spring effect counter acts changes in fluid mass, the absorber can be used as a water reservoir. However, early peaks in the response may require active control for their proper reduction. Consequently, such a hybrid absorber is discussed with gas injection from a standby high-pressure vessel, making it independent from any public energy transmission line.

1 Introduction

It is the horizontal component of the strong motion phase of an earthquake that shakes and possibly endangers regular buildings. To reduce the vulnerability of buildings with fundamental frequency higher than about 1 Hz (or buildings with

F. Ziegler (✉)
Vienna University of Technology, Vienna A-1040/E2063, Austria
e-mail: franz.ziegler@tuwien.ac.at

B. Khalid
Directorate of Civil Engineering and Design, Islamabad, Pakistan
e-mail: basitdotcom@hotmail.com

H. Irschik et al. (eds.), *Advanced Dynamics and Model-Based Control of Structures and Machines*, DOI 10.1007/978-3-7091-0797-3_26,

less than 10 stories), base isolation is best suited. Currently, a low pass filter is put between the building and its foundation mainly in the form of a shear spring consisting of rubber sheets reinforced by placing them between steel plates. The resulting fundamental mode should have a frequency below about 0.5 Hz. In that case the required damping of the resulting nearly rigid body mode of the base isolated building is supplied by the plastic deformation of a lead core placed in the center of the isolation element. This element has a rather large ratio of the vertical to the horizontal stiffness; the latter in a rough approximation is given in terms of the rubber shear modulus by $k_{1h} = GA/ns_1$, where A and s_1 are respectively, the cross-sectional area and the thickness of the n-times repeated rubber sheets. There are lifetime problems encountered, mainly caused by the increase of temperature due to plastic deformation of the lead core within the first occurrence of the earthquake or during the aftershocks. Consequently and avoiding any such lifetime problems, in this paper we describe a novel base isolation element consisting of an axially prestressed helical steel spring with maximized axial to horizontal (shear) stiffness combined with a pinned-pinned column (upright pendulum) carrying some portion of the dead weight, [Khalid B. Control of seismically forced vibrations of asymmetric buildings by means of a novel base isolation system. TU-Wien, www.tuwien.ac.at, Dissertation 2010]. Damping of the resulting fundamental modes after base isolation is supplied by the separate optimal arrangements of low cost Tuned Liquid Column Gas Dampers (TLCGD) in the building's basement. Such a system of base isolation has hardly any maintenance costs and has a lifetime comparable to that of the building. Early peaks in the response records are hardly affected by passive damping and their reduction requires active control. By means of controlled gas injection from a standby high-pressure vessel with subsequent removal the TLCGD becomes a hybrid absorber, an ATLCGD as outlined in this paper. The state-of-the-art of optimizing the two parameters, frequency and damping, of passively acting absorbers in the form of Tuned Mechanical Dampers (TMD) is based on Den Hartog [1], whereas that of active structural control up to 1990 is available in [2].

2 Equation of motion of the base isolated building

Due to the low pass filter placed between the building and its foundation, the horizontal deformations caused by the horizontal component of an earthquake are concentrated in the isolation elements and the resulting (three) low frequency fundamental modes can be considered as the rigid body motions of the building. The vertical motion of the building is a combination of the vertical component of the seismic forcing (no soil-structure interaction is considered) with a guided component from the horizontal motion. Consequently, for setting up the equations of horizontal motion by means of conservation of momentum and of angular momentum about the vertical axis, we need to know the total mass M_S, the moment of inertia $I_x = M_S r_S^2$ and the location of the center of mass C_M in the plan of the building. Hence, three equations result in matrix form, the mass matrix is diagonal,

rotation about x-axis $u_T = r_S\theta$, a single point seismic excitation by ground acceleration $a_g(t)$ with angle of incidence α against y-axis is understood,

$$\mathbf{M}\ddot{\mathbf{x}} + \mathbf{K}\mathbf{x} = -\mathbf{M}a_g\mathbf{e}_g + \mathbf{F}\,,\ \mathbf{x}^T = \begin{bmatrix} v_M\ w_M\ u_T \end{bmatrix}\,,\ \mathbf{e}_g^T = \begin{bmatrix} \cos\alpha\ \sin\alpha\ 0 \end{bmatrix}. \quad (1)$$

The generalized control force vector $\mathbf{F}$ supplies the action of the TLCGDs. Substructure synthesis of (26.1) with TMDs of the spring-mass-dashpot type would be exact, however that with equivalent TLCGDs becomes approximate since the (small amount) of dead fluid mass changes slightly the location of the center of mass, the mass matrix and thus the main system considered in (26.1). Extremely light structural damping of the low pass filter (not considered in (26.1)) will be added to the modally projected equations, Since we can assume isotropic horizontal stiffness of each of the base isolation elements, k_{1h}, and some equal spacing along the perimeter of the carrying walls of the building, the symmetric 3×3 stiffness matrix $\mathbf{K}$ of the base isolation is easily set up, e.g. by the direct stiffness matrix method. Considering a number N of such base isolation elements, the resulting horizontal stiffness is apparent in the two diagonal stiffness elements, $k_{11} = k_{22} = Nk_{1h}$. Thus, the required period renders a first condition to be met in the design of the base isolation by choosing the resulting stiffness,

$$T = 2\pi\sqrt{M_S/Nk_{1h}} \geq 2s. \quad (2)$$

Considering the simple rectangular plan $a \times b$ of an asymmetric building with an intermediate carrying wall and a symmetric distribution of the isotropic base elements, such that their number is $N = 2N_a + 3N_b$, the third diagonal component and the off-diagonal elements in the stiffness matrix become, Fig. 1,

$$\begin{aligned} k_{33} = \frac{Nk_{1h}}{r_S^2}\Bigg[& y_M^2 + z_M^2 + \frac{N_b}{2N}\left(a^2 + 2y_{wall}^2 - 4y_{wall}y_M\right) \\ & + \frac{N_a}{2N}b^2 + \frac{3}{N}\sum_{i=1}^{N_b} z_i^2 + \frac{2}{N}\sum_{i=1}^{N_a} y_i^2 \Bigg], \end{aligned}$$

$$k_{13} = k_{31} = Nk_{1h}\frac{z_M}{r_S}\,,\ k_{32} = k_{23} = Nk_{1h}\frac{y_M}{r_S}\left(\frac{N_b y_{wall}}{Ny_M} - 1\right). \quad (3)$$

2.1 *Modal analysis of the base isolated building*

Solving the linear eigen-value problem of (26.1), e.g., by calling the tool *eig* in Matlab, renders the set of three orthonormalized modal vectors and the associated natural frequencies, say around 0.5 Hz. The placement of the modally tuned

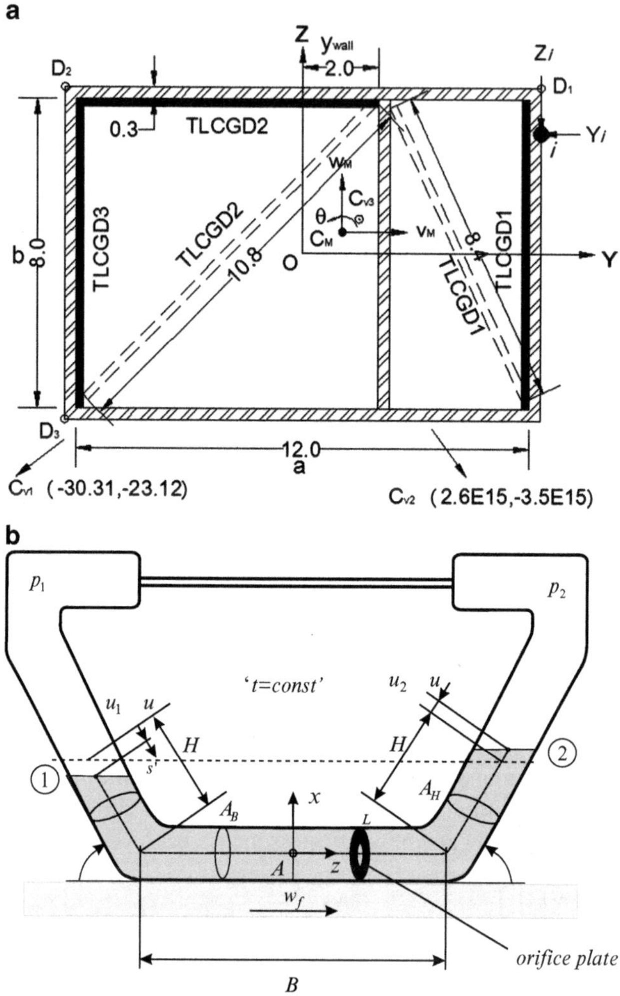

Fig. 1 (**a**) Plan of an asymmetric building. Spring force of a base isolation element at position i. Sub-optimal placements of the three TLCGDs (two alternatives) relative to C_{vj}. (**b**) Symmetric TLCGD. Framed piping system, sealed: one-sided gas-volume (equilibrium) $V_a = A_H H_a$

TLCGDs becomes optimal with normal distance to the modal centers of velocity maximum: centers far outside of the building plan indicate dominating translational motion in that mode, whereas dominating modal rotational vibrations render a modal center within the building plan. For small displacements, 3×1 modal vectors $\boldsymbol{\phi}_j$ determine the modal centers of velocity C_{vj},

$$y_{vj} = y_M - r_S \phi_{j2}/\phi_{j3}\,,\; z_{vj} = z_M + r_S \phi_{j1}/\phi_{j3}\,,\; \phi_{j3} \neq 0, \tag{4}$$

Ziegler [3], p. 13, see Fig. 1a. For the base isolated illustrative example, Fig. 1, with total mass $M_S \approx 245{,}000\,\mathrm{kg}$ and $r_S = 4.65\,\mathrm{m}$, the natural modes are given by

$$\begin{bmatrix} f_1 = 0.494 \\ f_2 = 0.500 \\ f_3 = 0.818 \end{bmatrix} Hz, \ \boldsymbol{\Phi} = 10^{-3} \begin{bmatrix} -1.211 & 1.617 & 0.143 \\ 1.606 & 1.220 & 0.190 \\ 0.238 & -2.15 \times 10^{-15} & 2.011 \end{bmatrix}. \tag{5}$$

The modal centers of velocity are located as indicated in Fig. 1a.

3 Design of the novel base isolation element

In Fig. 2 the base isolation element is sketched: with respect to the proper selection of the helical steel spring, we note the design equations for its vertical (pre-stress) and horizontal (shear) stiffness, and their ratio to be maximized within proper design limits, n_t is the number of turns,

$$k_{1v} = Ed\frac{(d/2D)^3}{2(1+\nu)n_t}, \ k_{1h} = Ed\frac{(d/2D)^3}{n_t\,[1+(4/3)(2+\nu)(l/D)^2]},$$

$$\lambda_{vh} = \frac{k_{1v}}{k_{1h}} = \frac{1+(4/3)(2+\nu)(l/D)^2}{2(1+\nu)} \Rightarrow \max. \tag{6}$$

When considering the static stability of the pivoted upright pendulum supported by the horizontal stiffness of the helical spring, Fig. 2a, only a fraction of the dead weight of the building (well below of the critical buckling load $k_{1h}l$, [3], p. 519), is carried by the column and thus axial prestressing $k_{1v}l(l_0/l - 1) > 0$ of the spring is required. For a number N of base isolation elements, the condition follows,

$$M_S g/Nk_{1h}l = \lambda + \lambda_{vh}(l_0/l - 1), \ \lambda < 1 \Rightarrow \lambda = 0.49, \text{ in [7]}. \tag{7}$$

Thus, (26.2) and (26.7) determine horizontal stiffness and the required number N with length l referred to the isolation gap between building and foundation selected.

4 Effective modal damping by the TLCGD

Hochrainer has invented the gas spring effect for easy frequency tuning of the TLCGD, Fig. 1b, see [4]. The equation of the relative fluid flow in the TLCGD, when installed in the basement of the building and with the trace oriented in a general direction, angle γ, is derived by integrating along the instant relative stream line to render a generalized non-stationary Bernoulli equation, [3], page 497 and in

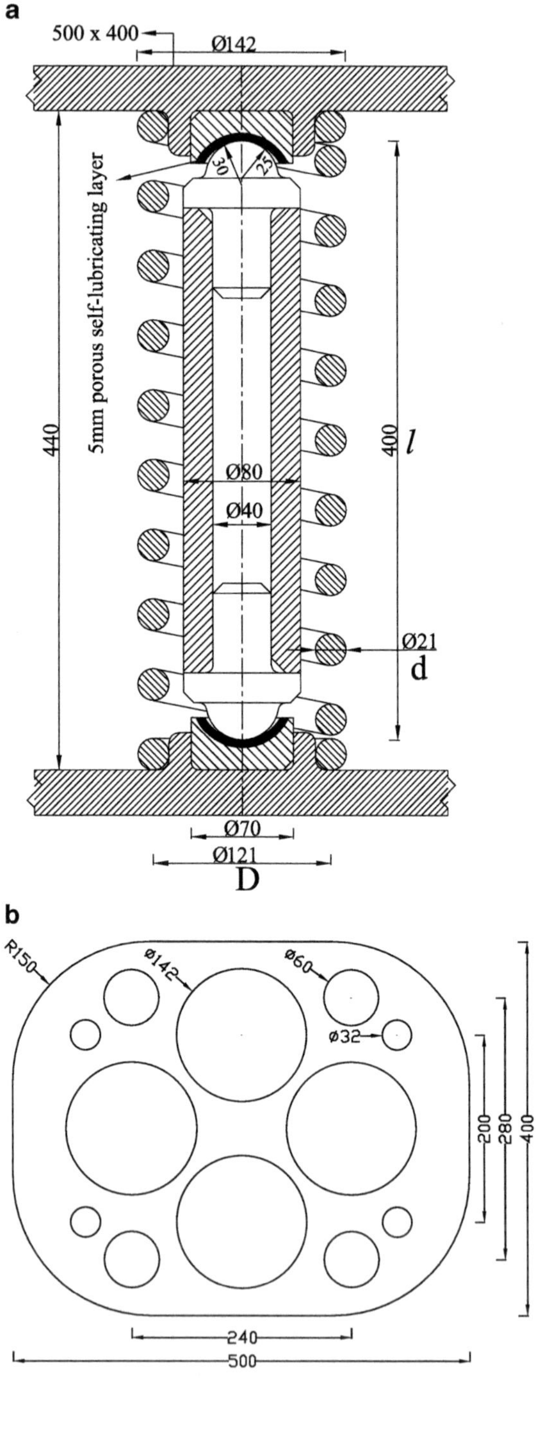

Fig. 2 (**a**) Novel base isolation element consisting of an axially prestressed helical steel spring and a pinned–pinned column in spherical bearings. (**b**) Schematic plan of a four-spring-pendulum unit with four-safety steel columns ($\Phi 60$) with axial clearance, maximum tilting angle $|\varphi| < 15°$

connection with the effective damping of multi-purpose, asymmetric buildings, [5]; $u_1 = u_2 = u(t)$ is the fluid stroke. For TLCGD numbered j,

$$\ddot{u}_j + 2\zeta_{Aj}\omega_{Aj}\dot{u}_j + \omega_{Aj}^2 u_j = -\kappa_j \left\{ \begin{array}{l} \left[a_g \cos\alpha + \ddot{v}_M - \ddot{u}_T(z_{Aj} - z_M)/r_S\right]\cos\gamma_j \\ + \left[a_g \sin\alpha + \ddot{w}_M + \ddot{u}_T(y_{Aj} - y_M)/r_S\right]\sin\gamma_j \end{array} \right\}$$

$$\kappa = \frac{B + 2H\cos\beta}{L_{eff}}, \quad L_{eff} = \frac{A_H}{A_B}B + 2H, \text{ all dimensions with number } j. \tag{8}$$

Considering the absorber damping sufficiently high, parametric excitation by both, the vertical component of base excitation and the rotation about the vertical x-axis becomes negligible and (26.8) results. The viscous damping in (26.8) represents the equivalently linearized, experimentally observed mean turbulent damping $\delta_L\,|\dot{u}|\,\dot{u}$: Hence, $\zeta_A = (4/3\pi)\delta_L \max|u|$. The linearized (26.8) enters the modal tuning process that, by analogy, can be based on the Den Hartog optimization of the equivalent TMD; see again [4] for details. Thus ζ_A is inserted with its optimal value. The absorber frequency in (26.8) takes on an optimal value (result of tuning) and should be set equal to the natural frequency of small vibrations of a mathematical pendulum, to render the most convenient design formula, defining the required equilibrium gas pressure, $np_0 = \rho g h_0$, $1 \leq n \leq 1.4$, in relation to the gas volume $V_a = A_H H_a$ required by the condition for linearized gas compression, $\max|u| \leq H_a/3$, number j understood, for a severe limit of f_A see [6],

$$f_A = \sqrt{\frac{g/\pi^2}{4(L_{eff}/2)}\left(\sin\beta + \frac{h_0}{H_a}\right)} = \sqrt{\frac{g/\pi^2}{4L_0}} \Rightarrow \frac{h_0}{H_a} = \frac{L_{eff}}{2L_0} - \sin\beta. \tag{9}$$

4.1 *Control forces for TLCGD acting in the basement of the base isolated rigid building, j understood*

Considering the absolute horizontal acceleration of the center of mass C_f of the displaced fluid column, conservation of both, momentum and angular momentum renders, the generalized control force components are listed in their linearized form to enter the approximate substructure synthesis with (26.1), [5],

$$\begin{aligned} F_y &= m_f\left[a_g\cos\alpha + \ddot{v}_M - \ddot{u}_T(z_A - z_M)/r_S\right] + \bar{\kappa} m_f \ddot{u}\cos\gamma\,,\ \bar{\kappa} = \kappa L_{eff}/L_1, \\ F_z &= m_f\left[a_g\sin\alpha + \ddot{w}_M + \ddot{u}_T(y_A - y_M)/r_S\right] + \bar{\kappa} m_f \ddot{u}\sin\gamma\,,\ m_f = \rho A_H L_1, \\ M_x &= m_f \bar{\kappa}_3 \ddot{u}_T H^2/r_S - F_y(z_A - z_M) + F_z(y_A - y_M), \\ \bar{\kappa}_3 &= \{[1 + (A_B/3A_H)(B/2H)](B/2H) + [1 + (2H/3B)\cos\beta]\cos\beta\}(B/L_1). \end{aligned} \tag{10}$$

4.2 Modal tuning

Equation (26.1) is modally expanded, $\mathbf{x} = \sum_{j=1}^{3} q_j(t)\boldsymbol{\phi}_j$, however, the modal coupling present in the generalized control forces (26.10) is neglected for the sake of modal tuning of the selected TLCGD by means of Den Hartogs optimization formulas. Considering this modal coupling in fine-tuning using optimizing tools in the state space can subsequently make a correction. The approximating result with (26.8) substituted for the isolated j-th mode is a 2 D.O.F. coupled system $[q_j\ u_j]^T$, the extremely light damping in the base isolation is commonly assumed constant, $\zeta_{Sj} = \zeta_S \approx 0.05$. The transformation of the optimal equivalent TMD parameters [1], is deduced by comparing two sets of equations, [4] and [6], a star refers to the TMD,

$$\delta_{jopt} = \frac{f_{Aj,opt}}{f_{Sj}} = \frac{\delta^*_{jopt}}{\sqrt{1+\mu_j\left[1-\kappa\bar{\kappa}(V_j^*/V_j)^2\right]}}, \quad \zeta_{Aj} = \zeta^*_{Aj}. \tag{11}$$

The dead fluid mass is apparent in the denominator in (26.11).

4.3 Fine tuning of the coupled set of TLCGDs in the state space

Full modal expansion of (26.1) and (26.10), (26.8) substituted, yields six coupled, still approximate modal equations. In the state space domain, the 12×1 state hyper vector for time harmonic excitation becomes,

$$\mathbf{z}_0(\alpha,\omega) = \left[i\omega\mathbf{I} - (\mathbf{A}+\mathbf{BR})\right]^{-1}\mathbf{E}_g\mathbf{e}_g a_0\,, \ \mathbf{z}_{0j} = \left[q_{0j}\ u_{0j}\ i\omega q_{0j}\ i\omega u_{0j}\right]^T. \tag{12}$$

The optimal parameters, contained in $\mathbf{R}$, are calculated by minimizing a performance index, e.g., considering the minimum of the area under the frequency response function of the main system, - e.g., by Matlab tool *fminsearch*, see [4,5].

4.4 Selected simulation results

Considering the base isolated single storey asymmetric building of Fig. 1a, forced by the El Centro seismogram, scaled to 0.32 g, convincingly approve the separated action of the low pass filter and of the three fine tuned TLCGDs, Fig. 3.

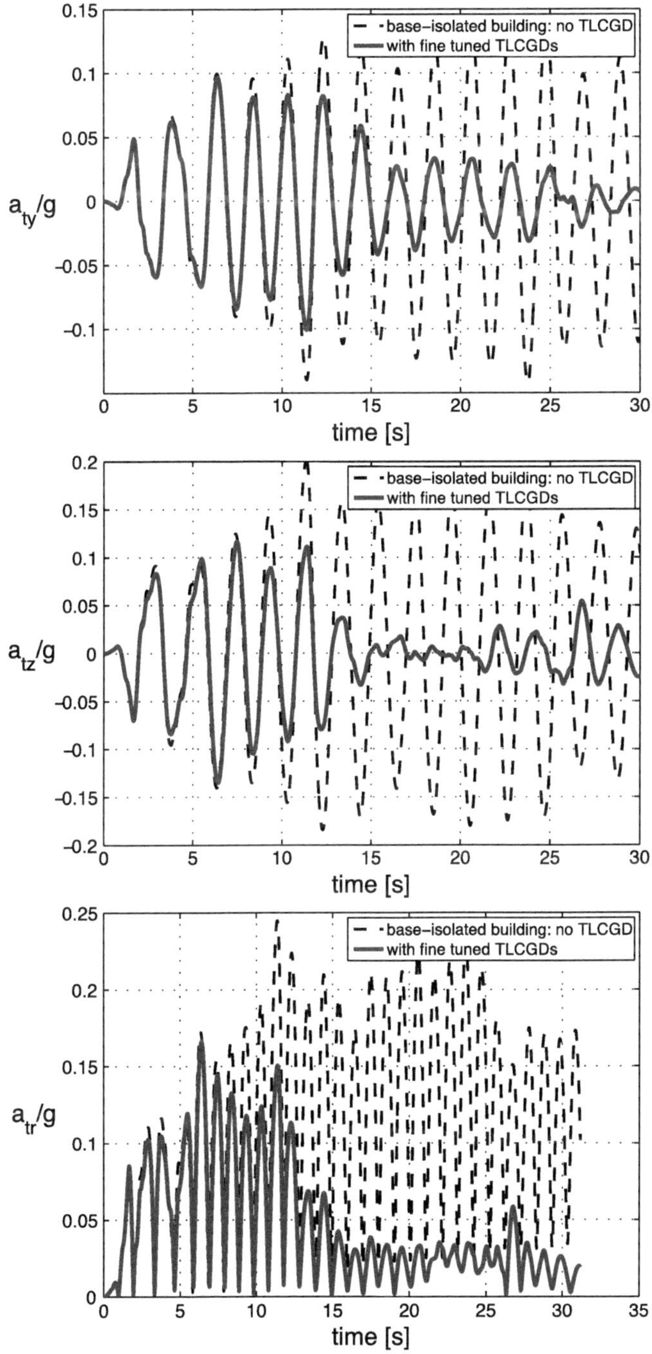

Fig. 3 Total horizontal acceleration of C_M and the resultant a_{tr}. El Centro, $\alpha_{cr} = 125°$

5 Active control rendering a hybrid ATLCGD

Early acceleration peaks in the response of the base isolated building are hardly affected by passive (effective) damping, Fig. 3. Adding a standby high pressure gas vessel to Fig. 1b, providing a few, bang-bang controlled gas pressure shots (the amount of gas must be controlled removed) $\Delta \bar{p}_a = -\Delta \bar{p}_{a,max} \mathrm{sign}(\mathbf{S}^{-1}\mathbf{E}_a^T \mathbf{P}\mathbf{z}_S)$, renders the ATLCGD hybrid and reduces such early peaks, [4],[2].

6 Conclusion

Simulations approve the novel base isolation with effective damping by TLCGDs. Steel springs in shear provide the horizontal stiffness with long lifetime. The pivoted column contacts self-lubricating spherical bearings to avoid corrosion.

7 Note added in proof

The novel base isolation system is complemented by separately acting "sliding elements" that provide limited static friction, however, without continuous energy dissipation in abrasive dry friction. For design details see Khalid [7].

References

1. Den Hartog, J.P.: Mechanical Vibrations, Reprint 4th edn. McGraw-Hill, New York (1956)
2. Soong, T.T.: Active Structural Control: Theory and Practice. Wiley, New York (1990)
3. Ziegler, F.: Mechanics of Solids and Fluids. Corr. 2nd edn. Springer, New York (1998)
4. Hochrainer, M.J., Ziegler, F.: Control of tall building vibrations by sealed tuned liquid column dampers. Struct. Control Health Monit. **13**, 980–1002 (2006)
5. Fu Chuan, Ziegler F.: Vibration prone multi-purpose buildings and towers effectively damped by tuned liquid column gas dampers. Asian J. Civil Eng. **10**(1) 21–56 (2009)
6. Ziegler F.: Special design of tuned liquid column-gas dampers for the control of spatial structural vibrations. Acta Mech. **201**, 249–267 (2008)
7. Khalid B.: Control of seismically forced vibrations of asymmetric buildings by means of a novel base isolation system. Dissertation, Vienna University of Technology (2010). http://www.ub.tuwien.ac.at/diss/AC07809200.pdf

Index

H. Irschik et al. (eds.), *Advanced Dynamics and Model-Based Control of Structures and Machines*, DOI 10.1007/978-3-7091-0797-3,

GPSR Compliance

The European Union's (EU) General Product Safety Regulation (GPSR) is a set of rules that requires consumer products to be safe and our obligations to ensure this.

If you have any concerns about our products, you can contact us on ProductSafety@springernature.com

In case Publisher is established outside the EU, the EU authorized representative is:

Springer Nature Customer Service Center GmbH
Europaplatz 3
69115 Heidelberg, Germany

Batch number: 10577866

Printed by Printforce, the Netherlands